Application of Stable Isotopes and Tritium in Hydrology

Application of Stable Isotopes and Tritium in Hydrology

Editors

Ines Krajcar Bronić
Jadranka Barešić

MDPI • Basel • Beijing • Wuhan • Barcelona • Belgrade • Manchester • Tokyo • Cluj • Tianjin

Editors
Ines Krajcar Bronić
Ruđer Bošković Institute
Croatia

Jadranka Barešić
Ruđer Bošković Institute
Croatia

Editorial Office
MDPI
St. Alban-Anlage 66
4052 Basel, Switzerland

This is a reprint of articles from the Special Issue published online in the open access journal *Water* (ISSN 2073-4441) (available at: https://www.mdpi.com/journal/water/special_issues/Stable_Isotopes_Tritium).

For citation purposes, cite each article independently as indicated on the article page online and as indicated below:

LastName, A.A.; LastName, B.B.; LastName, C.C. Article Title. *Journal Name* **Year**, *Volume Number*, Page Range.

ISBN 978-3-0365-0910-5 (Hbk)
ISBN 978-3-0365-0911-2 (PDF)

Cover image courtesy of Jadranka Barešić.

Contents

About the Editors

Ines Krajcar Bronić obtained her Doctor of Science (Physics) degree in 1993, University of Zagreb, Croatia, and joined the Ruđer Bošković Institute, Zagreb, Croatia, in 1982. Since 2009, she has held a senior scientist—tenure position; she is currently the Head of the Laboratory for Low-level Radioactivities. She has research experience in the interaction of low-energy radiation with gases, basic physical data used in radiotherapy, and her current research is devoted to application of isotope data (^{3}H, ^{14}C, ^{2}H, ^{13}C, ^{18}O) in interdisciplinary research fields, such as the chronology of carbonates in karst, environmental studies, climate change and the Anthropocene, radiocarbon dating of various cultural heritage objects. She was an Alexander-von-Humboldt fellow and Invited Research Fellow of the Japanese Society for Promotion of Science; Coordinator of FP6 AMS-^{14}C project "Preparation of carbon samples for ^{14}C dating by the AMS technique"; principal investigator of Croatian Science Foundation project "Reconstruction of the Quaternary environment in Croatia using isotope methods" and several bilateral projects. She is the Director of two IAEA Training Courses on Dating Techniques (2008, 2013).

Jadranka Barešić is a senior research associate at the Ruđer Bošković Institute. She obtained her PhD in chemistry in 2009, at the Faculty of Chemical Technology and Engineering, University of Zagreb, with the thesis "Application of isotopic and geochemical methods in monitoring of global and local changes in ecological system of Plitvice Lakes". Her main scientific interest is the application of radioactive (^{14}C and ^{3}H) and stable (^{2}H, ^{18}O, ^{13}C) isotopes in multidisciplinary research: heritage sciences, (paleo)environment, hydrogeology and geochemistry with emphasis on the Croatian karst. She is involved in the development and implementation of new techniques for sample preparation and measurement: benzene synthesis and measurement of ^{14}C activity by liquid scintillation spectrometry, sample preparation for measurement of ^{14}C by accelerator mass spectrometry (AMS) and electrolytic enrichment of water samples for ^{3}H measurements. She participated in many international and national projects, and recently, her main point of interest is climate change and the study of the Anthropocene.

Preface to "Application of Stable Isotopes and Tritium in Hydrology"

Studies of groundwater recharge and infiltration conditions, the mechanisms of mass transfer, and mixing of waters with different compositions and ages are among the basic problems of geology and hydrogeology. Isotopes of hydrogen (^{3}H, ^{2}H) and oxygen (^{18}O) are perfect candidates for groundwater tracers because they are constituent elements of the water molecule. Knowledge on the isotopic composition (Δ^{18}O, Δ^2H and ^{3}H activity concentration) in surface and groundwater has found wide application in hydrology, as well as environmental studies in hydro-ecological systems. The isotope techniques have proven to be effective tools for providing information that could not be obtained by any other conventional means.

This Special Issue (SI) of the MDPI journal Water addresses the current state-of-the-art applications of stable isotopes and tritium in studies of hydrological process and the whole water cycle. The SI comprises six papers that passed the rigorous review by at least two recognized international reviewers and extensive editorial checks. The compiled papers show a wide variety of topics: isotope distribution in precipitation and groundwater on regional scales, the relationship between isotopes in precipitation and ground and drip water in a karst area, effects of evapotranspiration, and farming activities on water isotope composition. Three out of six papers describe isotope studies in a karst area of Croatia, while the other papers consider the Iberian Peninsula, the United States of America, and China. The papers describe isotope applications in studies performed locally or regionally, but the conclusions obtained may be valid worldwide. Precipitation, groundwater, and surface waters belong to classical water bodies, while evapotranspiration, effects of farming, and drip water in karst caves seldom present applications of water isotopes.

There is still a lot of work to be performed jointly by the authors, the reviewers and the editors of the journal or Special Issues to enable a flawless transfer of ideas and acquired knowledge to the interested readership. The availability of new techniques and instruments that enable the fast measurement of a large number of small-sized samples at a reasonably good precision and a relatively low price will certainly increase the number of publications within the topics in isotope hydrology. We are looking forward to seeing new applications of the powerful isotope techniques.

Ines Krajcar Bronić, Jadranka Barešić
Editors

Editorial

Application of Stable Isotopes and Tritium in Hydrology

Ines Krajcar Bronić * and Jadranka Barešić

Laboratory for Low-Level Radioactivities, Division of Experimental Physics, Ruđer Bošković Institute, 10000 Zagreb, Croatia; jbaresic@irb.hr
* Correspondence: krajcar@irb.hr

Academic Editor: Y. Jun Xu
Received: 28 January 2021; Accepted: 3 February 2021; Published: 7 February 2021

The availability and quality of freshwater currently represent one of the great issues for mankind. Increasing demand for freshwater (both ground and surface water), environmental pollution, and climate change affects the water quality and changes its dynamics. Groundwater recharge and infiltration conditions, the mechanisms of mass transfer, and the mixing of water with different compositions and ages are among the basic problems of hydrology and hydrogeology that have been simultaneously tackled by both conventional hydrological methods and by isotope methods. Isotope studies applied to a wide spectrum of hydrological problems related to both surface and groundwater resources as well as environmental studies in hydro-ecological systems present today an established scientific discipline, often referred to as "Isotope Hydrology". Isotope hydrology techniques have proved to be effective tools for solving many critical hydrological problems and processes and in many cases, provide information that could not be obtained by conventional means only.

Isotopes of hydrogen (^{3}H, ^{2}H) and oxygen (^{18}O) are perfect candidates for groundwater tracers because they are constituent elements of the water molecule. Among them, ^{3}H (tritium) is the best candidate for the determination of groundwater age, if we assume that groundwater "age" is the travel time between recharge and discharge. Knowledge of the isotopic composition (δ^{18}O, δ^2H, including the derived quantity deuterium excess (*d*-excess), and ^{3}H activity concentration) in surface and groundwater and its relation to the isotopic composition of precipitation have found wide application in hydrology, such as the determination of the mean residence time of water in aquifers, the determination of recharge areas, the interconnections between aquifers, the origin of groundwater, and the dynamics of processes in surface waters.

A considerable number of studies on the use of isotope composition of various types of water have been published recently. In addition to the growing awareness of the usefulness of isotope studies in solving various hydrological problems, one of the main reasons for the increased number of isotope studies is the development of new measurement techniques [1] that enable fast measurement of a large number of small-sized samples at reasonably good precision and relatively low price.

This Special Issue (SI) of the MDPI journal *Water*, entitled "Application of stable isotopes and tritium in hydrology" addresses the current state-of-the-art applications of stable isotopes and tritium in studies of hydrological processes in various compartments of the water cycle. It can be considered as a sequel of the previous SI, "Use of Water Stable Isotopes in Hydrological Process" [2]. However, this SI is the only (closed) special issue of the journal *Water*—Hydrology and Hydrogeology Section that in the title explicitly mentions "tritium", the radioactive isotope of hydrogen of both natural (cosmogenic) and anthropogenic origin ([3] and references therein). The massive injection of anthropogenic tritium from atmospheric weapon tests in the 1950s and 1960s caused an almost 1000-fold increase in tritium activity concentration in precipitation, known as the "bomb peak". As precipitation is the ultimate source of groundwater, the bomb peak was quickly spread over the hydrosphere and the importance of tritium for studies of hydrological processes has been recognized by numerous researchers. In the last decades of the 20th century, tritium has been widely used to obtain time scales for physical processes in various hydrosphere compartments, mostly for dating modern water and distinguishing modern recharges from water older than about 50 years. After the cessation of atmospheric nuclear tests, a gradual decrease in tritium activity concentration in precipitation was observed worldwide until approximately the mid-1990s. After that, the mean annual tritium activity concentration in precipitation is almost constant approaching the pre-bomb tritium level and thus the present scientific value of tritium for hydrological applications has significantly declined [3].

This SI was announced and open for submission in November 2019, with a deadline of 24 May 2020. However, due to the pandemic and lockdown over most of the world in spring 2020, the deadline was extended to the end of July 2020. A total of nine papers were submitted for publication and they have undergone a scrutinized review process. Eventually, six papers passed the rigorous review by at least two recognized international reviewers and extensive editorial checks. The papers compiled in this SI show a wide variety of topics: isotope distribution in precipitation [4] and groundwater on regional scales [5], the relation between isotopes in precipitation and ground and drip water in a karst area [6,7], effects of evapotranspiration [8], and farming activities [9] on water isotope composition. Three out of six papers describe isotope studies in a karst area of Croatia [5–7], while the other papers consider the Iberian Peninsula [4], the United States of America [8], and China [9]. In what follows we shortly describe the most important aspects of the papers from this compilation and the main results presented.

Hatvani et al. [4] explored the spatial variation of oxygen and hydrogen stable isotope composition (δ_p) and d-excess of precipitation across the Iberian Peninsula. The data from 24 monitoring stations of the Global Network of Isotopes in Precipitation (GNIP) for October 2002–September 2003 were used, while for October 2004–June 2006, the 13 GNIP stations were merged with 21 monitoring stations from a regional network in NW Iberia. The research aimed to look for features of spatial variability of the isotope composition of precipitation in the Iberian Peninsula, to determine the spatial representativity of the regional precipitation monitoring network, and to provide an improved estimation for the spatial distribution of precipitation water stable isotopes across the Iberian Peninsula. Spatial autocorrelation structure of monthly and amount weighted seasonal/annual mean δ_p values were modeled, and two isoscapes were derived for stable oxygen and hydrogen isotopes in precipitation with regression kriging. Only using the GNIP sampling network, no spatial autocorrelation structure of δ_p could have been determined due to the scarcity of the network. However, in the case of the merged and much denser GNIP and NW dataset, for δ_p a spatial sampling range of ~450 km in the planar distance (corresponding to ~340 km in geodetic distance) was determined. The range of δ_p, which also broadly corresponded to the range of the d-excess, probably referred to the spatially variable moisture contribution of the western Atlantic-dominated and eastern Mediterranean-dominated domains of the Iberian Peninsula. The estimation error of the presented Iberian precipitation isoscapes, both for oxygen and hydrogen, was smaller than the ones that were reported for the regional subset of one of the most widely used global models, suggesting that the current regional model provided a higher predictive power. The authors concluded that a sparser network (e.g., GNIP) might be representative and suitable for the Iberian Peninsula, but its verification could only be done with a denser network. The results encourage the development of precipitation stable isotope models at a sub-continental scale in further regions.

Brkić et al. [5] presented an overview of isotope investigations in both karst and alluvial aquifers in Croatia at approximately 100 sites from the period 1997–2014. Stable isotopes of oxygen ($\delta^{18}O$) and hydrogen (δ^2H) were used as indicators of the recharge condition, while tritium (^3H) was used to estimate an approximate mean groundwater age and for that the qualitative criteria were used to distinguish water recharged prior to the

1950s (<0.8 TU, sub-modern water), a mix of sub-modern and modern water (0.8 to about 4 TU), modern water (5 to 15 TU), waters containing "bomb" tritium (15–30 TU), and waters recharged in the 1960s to 1970s (>30 TU). The regional meteoric water lines (RMWL) were determined for the continental part ($\delta^2 H_{p,cont} = (7.400 \pm 0.005)\,\delta^{18}O + (4.1 \pm 0.5)$, $R^2 = 0.99$, $n = 524$) and the coastal part of Croatia ($\delta^2 H_{p,coast} = (7.00 \pm 0.08)\,\delta^{18}O + (4.4 \pm 0.5)$, $R^2 = 0.96$, $n = 655$). The composition of the stable isotopes of groundwater originated from recent precipitation and was described by regional groundwater lines (RGWL) that closely resembled the RMWLs: groundwater accumulated in the aquifers in the Pannonian (continental) part of Croatia can be described by $\delta^2 H_{g,cont} = 7.4\,\delta^{18}O + 5.5$, $n = 255$, $R^2 = 0.93$, and groundwater accumulated in the Dinaric karst of Croatia by $\delta^2 H_{g,coast} = 7.0\,\delta^{18}O + 4.2$, $n = 340$, $R^2 = 0.96$. The d-excess values of karst groundwater varied between 6‰ and 17‰ and increased with altitude; the catchment areas at relatively high altitudes and under the influence of the Mediterranean air masses showed the highest d-excess values, while the lowest d-excess values were determined in springs with catchment areas at relatively low altitudes and affected by the Atlantic air masses. The individual d-excess values of groundwater in the continental (Pannonian) region varied between 7.8‰ and 14.4‰. Higher values were again attributed to the influence of Mediterranean precipitating air masses. Most of the d-excess values in the groundwater samples in the Sava basin, Eastern Slavonia ranged from 10‰ to 12‰. Here the mean residence time of the groundwater was relatively long and, accordingly, the different individual precipitation contributions to groundwater were well mixed and their variations were dampened in the groundwater samples. Opposite to this, a wide range of d-excess values was found in the groundwater samples with relatively short mean residence time. The isotope content showed that the studied groundwater was mainly modern water making such water relatively vulnerable to the potential sources of environmental pollution. A mix of sub-modern and modern water was mostly accumulated in semi-confined porous aquifers in northern Croatia, deep carbonate aquifers, and (sub)thermal springs, which puts groundwater in these areas in a better position regarding vulnerability and pollution risks. This paper presents an important contribution to regional knowledge on groundwater hydrology obtained by applying tracers such as stable isotopes and tritium. The presented results may be useful for future studies on the isotopic composition of groundwater in Croatia and in nearby countries.

Although Brkić et al. [5] included about 100 sites from two regions in Croatia, they did not include the area of the Plitvice Lakes (PL) that presents a unique karst geomorphological system of 16 cascade flow-through lakes that are fed by three main springs and outflow to the Korana River. Krajcar Bronić et al. [6] presented various isotope studies of different types of water bodies (precipitation, groundwater, surface lake and river water, lake water from traps at certain depths) from the early period of isotope applications (since 1979) to the most recent ones (2018). The aim of the paper was to evaluate the most important hydrological inputs to the Plitvice Lakes and to detect possible effects of climate change on karst groundwater. The study included tritium, ^2H, and ^{18}O, and available climatological data (amount of precipitation, air temperature) in a search for evidence of climatological changes. An increase in the mean annual air temperature of 0.06 °C/year and the annual precipitation amount of 10 mm/year was observed for the PL area. The good correlation of the tritium activity concentration in the PL and Zagreb precipitation (discussed in detail in [3]) implied that the tritium data for Zagreb were applicable for the study of the PL area. The best local meteoric water line at PL was obtained by the reduced major axis regression (RMA) and precipitation-weighted ordinary least squares regression (PWLSR) approaches that gave also the best results for the closest precipitation stations Zagreb and Zadar: $\delta^2 H_{PWLSR} = (7.97 \pm 0.12)\,\delta^{18}O + (13.8 \pm 1.3)$, $n = 36$. The higher deuterium excess at PL (14.0 ± 2.2‰) than that at Zagreb (8.8 ± 0.8‰) in the same period reflected the higher altitude and influence of the Mediterranean precipitation at PL. The $\delta^2 H$ in precipitation ranged from -132.4‰ to -22.3‰ and $\delta^{18}O$ from -18.3‰ to -4.1‰. The much narrower ranges in the groundwater (<1‰ in $\delta^{18}O$, <10‰ in $\delta^2 H$) indicated good mixing of waters in the aquifers. The higher average $\delta^2 H$ in all three karst springs observed after 2003 was attributed to the increase in the mean air temperature. The mean $\delta^2 H$ and $\delta^{18}O$ values in the surface and lake water increased downstream due to the evaporation of surface waters. It was shown that the stable isotope composition of surface and lake waters reacted to extreme weather conditions (intense precipitation events and snow melting). Tritium activity concentration was used to determine the mean residence time (MRT) in three karst springs (range 2–4 years). Extreme hydrological conditions in 1983 (very low flow rate) and 1984 (high flow rate due to snow melting) enabled the determination of relative contributions of base-flow and precipitation in the three karst springs: the shorter the MRT, the higher proportions of precipitation. To conclude, the long-term and comprehensive isotope study of different water bodies in the area of the Plitvice Lakes can serve as an example of how the application of water isotopes (^{2}H, ^{18}O, ^{3}H) can help in the characterization of karst aquifers on regional and global scales.

Dinaric karst covers almost half of the Croatian territory [7], including islands and the Adriatic coast, high mountain regions, and part of central Croatia. The karst region is characterized by various forms of secondary carbonate sediments, like speleothem, tufa, or carbonaceous lake sediments. Speleothems deposited from cave drip waters retain in their calcite lattice isotopic records of past environmental changes and thus present the natural archives which confidentially record hydroclimate changes. Among other proxies, δ^{18}O is recognized as very useful for this purpose, but its accurate interpretation depends on understanding the relationship between precipitation and drip water δ^{18}O, a relationship controlled by climatic settings. Under equilibrium deposition conditions, speleothem δ^{18}O depends solely on cave air temperature and drip water δ^{18}O. In that case, the underground environmental changes respond to long-term climate changes and are recorded in the isotopic composition of slow-growing speleothems. Establishing cave hydroclimate monitoring is critical to assess variability in these two parameters. Surić et al. [7] analyzed isotope data of precipitation and drip water from 17 caves from different latitudes and altitudes in relatively small but diverse Croatian karst regions in order to distinguish the dominant influences. Stable isotope composition of precipitation collected at sites near the caves showed that local meteoric water lines (LMWL) lie between the Global MWL (δ^2H = 8 δ^{18}O + 10) and the Western Mediterranean MWL (δ^2H = 8 δ^{18}O + 13.7). Although slopes of the LMWLs were systematically lower (between 6.6 and 7.8) than the slope of the GMWL, a distinct difference was obtained between the northern continental stations (slope closer to 8) and maritime sites with lower slopes indicating enhanced secondary evaporation. Drip water δ^{18}O in colder caves generally resembled the amount-weighted mean of precipitation δ^{18}O. In these caves, in near-equilibrium calcite precipitation, speleothem δ^{18}O variations can reflect past meteoric precipitation and air temperature. At warmer sites, where evaporation played an important role, drip water δ^{18}O was usually more negative than the amount-weighted average δ^{18}O in precipitation. However, during glacial periods, today's "warm" sites were cold, changing the cave characteristics and precipitation δ^{18}O transmission patterns. Superimposed on these general settings, each cave has site-specific features, such as morphology (descending or ascending passages), altitude and infiltration elevation, (micro) location (rain shadow or seaward orientation), aquifer architecture (responsible for the drip water homogenization), and cave atmosphere (governing equilibrium or kinetic fractionation). This necessitates an individual approach to each cave and thorough monitoring for best comprehension. The authors stress that this approach should be the first and foremost guiding principle against which all actions need to be conducted worldwide in forthcoming studies of paleoclimate and paleo-environmental conditions based on cave speleothems.

Evapotranspiration (ET) is a process in the terrestrial hydrological cycle that accounts for evaporation (E) from the soil, open water, and canopy-intercepted water, and transpired (T) water from vegetation. ET constitutes a large percentage of the water cycle in most environments, and up to 95% in arid environments, being, therefore, an important water flux at a variety of spatial scales. Adkinson et al. [8] presented the potential of high-frequency partitioning ET into its constituent fluxes, T and E, which is important for understanding water use efficiency in forests and other ecosystems. Recent advancements in cavity ring-down spectrometers (CRDS) have made collecting high-resolution (temporal) water isotope data possible in remote locations, but this technology has rarely been utilized for partitioning ET in forests and other natural systems. To understand how the CRDS can be integrated with more traditional techniques, they combined stable isotope, eddy covariance, and sap flux techniques to partition ET in oak woodland in Texas, USA, using continuous water vapor CRDS measurements and monthly soil and twig samples processed using isotope ratio mass spectrometry (IRMS). They also wanted to compare the efficacy of δ^2H versus $\delta^{18}O$ within the stable isotope method for partitioning ET. The isotopic composition (δ^2H and $\delta^{18}O$) of atmospheric and precipitation samples fell closely along the global meteoric water line (GMWL), while the soil samples fell to the right of the GMWL, indicating preferential evaporation of lighter isotopologues of water and soil enrichment with the heavier isotopologues. These results indicated the evaporative fractionation of soil water and an appropriate degree of separation between the signals of twig and soil water isotope ratios. It was shown that average daytime vapor pressure deficit and soil moisture could successfully predict the relative isotopic compositions of soil and xylem water, respectively. Contrary to past studies, δ^2H and $\delta^{18}O$ performed similarly, indicating CRDS can increase the utility of $\delta^{18}O$ in stable isotope studies. However, they found a 41–49% overestimation of the contribution of T to ET when utilizing the stable isotope technique compared to traditional techniques (reduced to 4–12% when corrected for bias), suggesting there may be a systematic bias to the Craig-Gordon Model in natural systems. This study demonstrated the utility of using a combination of stable isotopes, sap flux, and eddy covariance techniques to partition ET in oak woodland and introduced a novel methodology using CRDS water vapor isotopes. However, the authors pointed out the need for further refinement of the methodology, particularly in natural systems, to reconcile potential biases inherent in this approach, and to test the proposed models under various natural conditions.

Landform changes caused by human activities can directly affect the recharge of groundwater and are reflected in the temporal and spatial changes in groundwater stable isotope composition. These changes are particularly evident in high-intensity farming areas. Together with recent climate changes, they may influence local and regional water resources management and sustainability of local life. Liu et al. [9] tested and analyzed groundwater stable isotope samples at different elevations of rice terraces in a typical agricultural watershed of the Hani Terraces, a World Heritage Cultural Landscape in southwest China. They analyzed the features of the elevation effect on the oxygen and hydrogen isotopic composition of groundwater across month and year; analyzed and identified the influencing mechanisms of land-use change caused by farming activities on the elevation gradient of groundwater isotopes from a spatial perspective; estimated the recharge cycle and period of regional groundwater from a temporal perspective using the deuterium excess, and deepened the understanding of how farming activities can improve groundwater recharge efficiency and maintain agricultural sustainability. They determined the characteristic variations and factors that influence the temporal and spatial effects on groundwater stable isotopes in the Hani Terraces, which are under the influence of high-intensity farming activities. The relation between δ^2H and $\delta^{18}O$ in groundwater was described as $\delta^2H = 4.98\,\delta^{18}O - 15.01$, $R^2 = 0.92$, $n = 144$, with no significant difference between the rainy and dry seasons, and with the slope lower than the LMWL slope of 8.31. The elevation gradients of $\delta^{18}O$ and δ^2H in groundwater were significantly increased due to farming activities when compared to the altitude gradients of precipitation. The values were -0.88‰ $(100\,\text{m})^{-1}$ and -4.5‰ $(100\,\text{m})^{-1}$, respectively, and they changed with time. The authors proposed using the elevation gradient of groundwater as a key index for assessing groundwater recharge. The groundwater circulation cycle was determined to be approximately three months, based on variations in the deuterium excess value of groundwater. The authors used the special temporal and spatial variation characteristics of the groundwater isotopes to evaluate the source and periodic changes of groundwater recharge. In addition, high-intensity rice farming activities, such as plowing every year from October to January, can increase the supply of terraced water to groundwater, thus ensuring the sustainability of rice cultivation in the terraces during the dry season. This demonstrates the role of human wisdom in the sustainable and benign transformation of surface cover and the regulation of groundwater circulation.

Finally, we can proudly conclude that the six scientific papers belonging to this SI showed a wide variety of isotope applications in various studies performed locally or regionally, but the conclusions obtained may be valid worldwide. Precipitation, groundwater, and surface waters belong to classical water bodies, while evapotranspiration, effects of farming, and drip water in karst caves present relatively seldom applications of water isotopes.

As authors, guest editors, and reviewers of various manuscripts submitted to *Water* and other MDPI journals, we consider that the following reflections (or "lessons learned") could be useful for future publications. The use of appropriate and unified terminology is a major issue to be resolved by the joint effort of authors, reviewers, and editors. Correct and clear language is also indispensable for the flawless transfer of ideas and acquired knowledge. Suitable use of statistics (both the choice of statistical methods and statistical expression of results with uncertainties and significance levels) should be a prerequisite for scientific conclusions based on a large number of individual data, as is often the case in long-term studies. Clear graphical presentations and appropriate presentation of data in tables must not be forgotten, as well as the presentation of raw data, usually as supplementary files, and proper referencing.

The Guest Editors are looking forward to the continuation of this SI through the new Special Issue "Isotope Fingerprints of Precipitation in Groundwater, Lakes and Rivers" which is open for submission until 30 October 2021. As has been mentioned many times, precipitation is the ultimate source of groundwater and surface water. Recent climate change has also caused variations in the amount of precipitation and their isotopic composition. Lack of precipitation can cause deterioration of surface water discharges and a decrease in groundwater levels which can lead to water scarcity for both human consumption and ecosystem needs. On the other hand, an extreme amount of precipitation can cause problems such as flooding. Variations in the isotopic composition in precipitation are reflected in the isotopic composition of groundwater and surface waters (rivers and lakes), and, therefore, the isotope composition of all three water body types will help detect any change in these inter-dependent systems caused by climate change, anthropogenic activities, or natural disasters such as volcanic eruption, regional fires, etc. This new SI of *Water* will focus on this relationship, not just taking into account ^{18}O and ^{2}H but all isotopes and hydrogeological parameters that can provide information about water resources.

Author Contributions: Conceptualization, I.K.B.; formal analysis, I.K.B. and J.B.; writing, reviewing and editing, I.K.B. and J.B. Both authors have read and agreed to the published version of the manuscript.

Funding: The Guest Editors' activities for this Special Issue received no external funding.

Acknowledgments: The Guest Editors express their gratitude to the Water MDPI in-house editors for their valuable help during the proceedings of the manuscripts. The Guest Editors also thank all the co-authors for their appreciated contributions and all the reviewers for their constructive comments and suggestions that greatly improved the outcome of the SI.

Conflicts of Interest: The authors declare no conflict of interest.

References

1. Vreča, P.; Kern, Z. Use of Water Isotopes in Hydrological Processes. *Water* **2020**, *12*, 2227. [CrossRef]
2. Vreča, P.; Kern, Z. (Eds.) *Use of Water Stable Isotopes in Hydrological Process, Special Issue*; MDPI: Basel, Switzerland, 2020; ISBN 978-3-03943-267-7. [CrossRef]
3. Krajcar Bronić, I.; Barešić, J.; Borković, D.; Sironić, A.; Mikelić, I.L.; Vreča, P. Long-Term Isotope Records of Precipitation in Zagreb, Croatia. *Water* **2020**, *12*, 226. [CrossRef]
4. Hatvani, I.G.; Erdélyi, D.; Vreča, P.; Kern, Z. Analysis of the Spatial Distribution of Stable Oxygen and Hydrogen Isotopes in Precipitation across the Iberian Peninsula. *Water* **2020**, *12*, 481. [CrossRef]

5. Brkić, Ž.; Kuhta, M.; Hunjak, T.; Larva, O. Regional Isotopic Signatures of Groundwater in Croatia. *Water* **2020**, *12*, 1983. [CrossRef]

6. Krajcar Bronić, I.; Barešić, J.; Sironić, A.; Mikelić, I.L.; Borković, D.; Horvatinčić, N.; Kovač, Z. Isotope Composition of Precipitation, Groundwater, and Surface and Lake Waters from the Plitvice Lakes, Croatia. *Water* **2020**, *12*, 2414. [CrossRef]

7. Surić, M.; Czuppon, G.; Lončarić, R.; Bočić, N.; Lončar, N.; Bajo, P.; Drysdale, R.N. Stable Isotope Hydrology of Cave Groundwater and Its Relevance for Speleothem-Based Paleoenvironmental Reconstruction in Croatia. *Water* **2020**, *12*, 2386. [CrossRef]

8. Adkison, C.; Cooper-Norris, C.; Patankar, R.; Moore, G.W. Using High-Frequency Water Vapor Isotopic Measurements as a Novel Method to Partition Daily Evapotranspiration in an Oak Woodland. *Water* **2020**, *12*, 2967. [CrossRef]

9. Liu, C.; Jiao, Y.; Zhao, D.; Ding, Y.; Liu, Z.; Xu, Q. Effects of Farming Activities on the Temporal and Spatial Changes of Hydrogen and Oxygen Isotopes Present in Groundwater in the Hani Rice Terraces, Southwest China. *Water* **2020**, *12*, 265. [CrossRef]

Article

Analysis of the Spatial Distribution of Stable Oxygen and Hydrogen Isotopes in Precipitation across the Iberian Peninsula

István Gábor Hatvani [1], Dániel Erdélyi [1,2], Polona Vreča [3] and Zoltán Kern [1,*

[1]   Institute for Geological and Geochemical Research, Research Centre for Astronomy and Earth Sciences, MTA Center for Excellence, Budaörsi út 45, H-1112 Budapest, Hungary; hatvaniig@gmail.com (I.G.H.); danderdelyi@gmail.com (D.E.)

[2]   Centre for Environmental Sciences; Department of Geology, Eötvös Loránd University, Pázmány Péter stny 1, H-1117 Budapest, Hungary

[3]   Department of Environmental Sciences, Jožef Stefan Institute, Jamova 39, 1000 Ljubljana, Slovenia; polona.vreca@ijs.si

*   Correspondence: zoltan.kern@gmail.com

Received: 15 November 2019; Accepted: 6 February 2020; Published: 11 February 2020

Abstract: The isotopic composition of precipitation provides insight into the origin of water vapor, and the conditions attained during condensation and precipitation. Thus, the spatial variation of oxygen and hydrogen stable isotope composition (δ_p) and d-excess of precipitation was explored across the Iberian Peninsula for October 2002–September 2003 with 24 monitoring stations of the Global Network of Isotopes in Precipitation (GNIP), and for October 2004–June 2006, in which 13 GNIP stations were merged with 21 monitoring stations from a regional network in NW Iberia. Spatial autocorrelation structure of monthly and amount weighted seasonal/annual mean δ_p values was modelled, and two isoscapes were derived for stable oxygen and hydrogen isotopes in precipitation with regression kriging. Only using the GNIP sampling network, no spatial autocorrelation structure of δ_p could have been determined due to the scarcity of the network. However, in the case of the merged GNIP and NW dataset, for δ_p a spatial sampling range of ~450 km in planar distance (corresponding to ~340 km in geodetic distance) was determined. The range of δ_p, which also broadly corresponds to the range of the d-excess, probably refers to the spatially variable moisture contribution of the western, Atlantic-dominated, and eastern, Mediterranean-dominated domain of the Iberian Peninsula. The estimation error of the presented Iberian precipitation isoscapes, both for oxygen and hydrogen, is smaller than the ones that were reported for the regional subset of one of the most widely used global model, suggesting that the current regional model provides a higher predictive power.

Keywords: geostatistics; isotope hydrometeorology; oxygen and hydrogen stable isotopes; isoscape; atmospheric moisture

---

1. Introduction

Since the discovery that isotopic composition of precipitation can give an insight into the origin of water vapor, the conditions attained during condensation and precipitation [1], stable water isotopes have become important natural tracers in the study of the water cycle [2,3]. Isotope ratios in precipitation mainly depend on geographic position and climate conditions [1,4]. In Europe, the continental, altitude, and latitude effects are the most important [5].

Our understanding of the behavior of water isotopes has rapidly improved thanks to recent advances in the measurement and modeling of stable water isotopes [6]. The spatial autocorrelation structure of the characteristics of isotopes is a feature of precipitation that has not been extensively studied, despite the fact that the geostatistical assessment of water stable isotopes in precipitation ($\delta^{18}O_p$, δ^2H_p) recorded at multiple monitoring stations can provide additional information that cannot be retrieved from single-station studies, and is therefore unavoidable in the exploration of the spatial representativeness of a regional monitoring network [7]. One reason for the scarcity of geostatistical studies on $\delta^{18}O_p$ and δ^2H_p records might be that a large number of stations in a relatively well populated network is required if a geostatistical evaluation is to be conducted.

Interpolated maps representing the global distribution of water stable isotopes in precipitation have been produced with geostatistical approaches of different complexity. These range from more simple models using common regression analysis (e.g., [8]) or geostatistical interpolation (e.g., [9,10]) to more complex regression-kriging approaches employing different auxiliary factors on regional scales [11–15] and on a global scale [16,17]. Isoscapes that are derived by complex geostatistical models will better approximate the actual variability of isotopes in precipitation, thus help researchers within the hydrological community and in other fields that leverage these products [3]. These products later on serve as benchmarks in many fields of science, e.g., hydrogeology [3,18], ecology [19], food source traceability [20], and forensic sciences studies [21]. However, a solid knowledge of the spatial autocorrelation/dependence structure of precipitation stable isotopes (δ_p) can be crucial to the derivation of precipitation isotopic landscapes (isoscapes), which are used to estimate local isotopic composition of precipitation as a function of observed local and/or extra local environmental variables across space and time while using gridded environmental data sets [22].

A global isotope-hydrometeorological monitoring network (GNIP—Global Network of Isotopes in Precipitation) was launched in the 1960s [4,23]. A European sub-region with one of the highest abundances of precipitation isotope monitoring stations can be found on the Iberian Peninsula (Figure 1). The first GNIP station to measure precipitation stable isotopes on the Peninsula was in Gibraltar, operating between October 1961 and February 1966, and later, from December 1972 onwards, providing continuous monthly data. The basic characteristics of precipitation stable isotopes have been assessed for both Spain [24] and Portugal [25], but a thorough geostatistical analysis of the whole peninsula has yet to be conducted. Recent results from large-scale moisture transport modeling presented spatiotemporally variable moisture contribution for the Iberian Peninsula [26,27]. However, former isotope hydrometeorological studies that were conducted on a sub-region and/or employing a limited number of stations, reported contradictory results in this respect. For instance, the springtime dominance of the Mediterranean-sourced precipitation was reported for southern and eastern Spain [28], while others conveyed that the Atlantic Ocean is the main vapor source over the Iberian Peninsula [29]. Assessing the spatial autocorrelation structure of precipitation stable isotope data retrieved from an extensive set of station record could provide a new perspective in this debate, since the isotopic composition of local precipitation is primarily controlled by regional scale processes [30].

Models and derived maps for the spatial distribution of precipitation water stable isotopes across the Iberian Peninsula are available as a part of global isoscapes [16,31] or were only developed for a sub-domain of the peninsula while using only a limited set of available measurements [29].

The aim of this research was to look for features of spatial variability of δ_p in the Iberian Peninsula, other than those related to well-known isotopic effects in an explorative approach. Specifically, to (i) determine the spatial representativity of its precipitation monitoring network and (ii) provide an improved estimation for the spatial distribution of precipitation water stable isotopes across the Iberian Peninsula.

2. Materials and Methods

2.1. Study Area

The Iberian Peninsula is located in the mid-latitudes of the Northern Hemisphere between the Mediterranean Sea and the Atlantic Ocean. Four major climate zones may be delineated on the peninsula (Figure 1A). The greater part might be categorized as warm temperate, with dry, hot, or cool summers (Köppen codes: Csa/Csb, respectively; Figure 1A), where dry hot summers are dominant (Csa;), and the value of δ_p had risen to around zero during the light summer rains (Gibraltar, Figure 1C). The values of d-excess (d = δ^2H – 8 × δ^{18}O, [1]) indicate an intra-annual trend with the minimum being reached in mid-summer.

Cool summers are more frequent (Csb) in the northwestern parts of the peninsula. The proximity of the ocean causes both milder winters and cooler summers, as well as provides an abundant moisture supply, giving a fair amount of annual precipitation evenly distributed over the year [32]. Precipitation δ^{18}O values are usually below −3‰, peaking in August with a dampened intra-annual trend characteristic of d-excess as well (e.g., La Coruña; Figure 1B). This was generally below 10‰.

Cold, semi-arid (BSk) conditions are prevalent in the more elevated central and southeastern regions of the peninsula. These result in the most characteristic intra-annual pattern in terms of stable isotopic features of precipitation. In Madrid, for instance, the seasonal amplitude of δ^{18}O$_p$ is ~12‰ and the maximum δ^{18}O$_p$ occurs in mid-summer (July) (Figure 1E). A similar δ^{18}O$_p$ pattern was reported for the years 2000–2002 [24]. Monthly d-excess values are characteristically below 10‰.

Humid warm summers (Cfb) characterize the northern sector of the peninsula, in the vicinity of the Pyrenees and along the Cantabrian coast. Here, precipitation δ^{18}O shows a minimum value in February and a maximum in May (e.g., Santander, Figure 1D). Deuterium-excess displays a relatively small degree of seasonal fluctuation, with monthly values characteristically above 10‰.

The most important moisture sources for the region are the tropical-subtropical North Atlantic corridor, the western Mediterranean basin, and recycled water from the Iberian land-mass, though to a varying extent intra-annually. In winter, the Atlantic serves as almost the exclusive moisture source in the western parts of the peninsula [26], and no influence from the Mediterranean moisture could be detected west from ~5°W longitude [27]. In summer, the western Mediterranean region provides almost twice as much moisture as North Atlantic, and it has a stronger effect in the eastern regions [26,27].

2.2. Used Materials

Precipitation oxygen and hydrogen stable isotope composition is conventionally expressed in δ values: the ratio between the heavy and light stable isotopes (i.e. ^{18}O/^{16}O or ^{2}H/^{1}H) in the sample when compared to the Vienna Standard Mean Ocean Water (V-SMOW), being expressed in per mill [33].

Monthly δ_p values were acquired from 32 Iberian stations of the Global Network of Isotopes in Precipitation (GNIP). The available data were unequally distributed in time and space (Figures 1 and 2). Historically, after Gibraltar, the second station was Faro, while the GNIP network on the peninsula became spatially extensive after May 1988, with 10 stations. Between 1992 and 1998, the number of simultaneously operating stations once again fell to ~5. The number of operating stations was usually ≥15 between 1998 and 2011, after which once more, only two stations, Gibraltar and Riells, were gathering samples [23]. The point at which the network was at its most extensive was in January 2003, when 24 monitoring sites were operating simultaneously (Figure 2). In the period assessed here (October 2002–September 2003), the same number of data were generally available for both water stable isotopes, although, in certain cases, the δ^{18}O records were more abundant (red line in Figure 2).

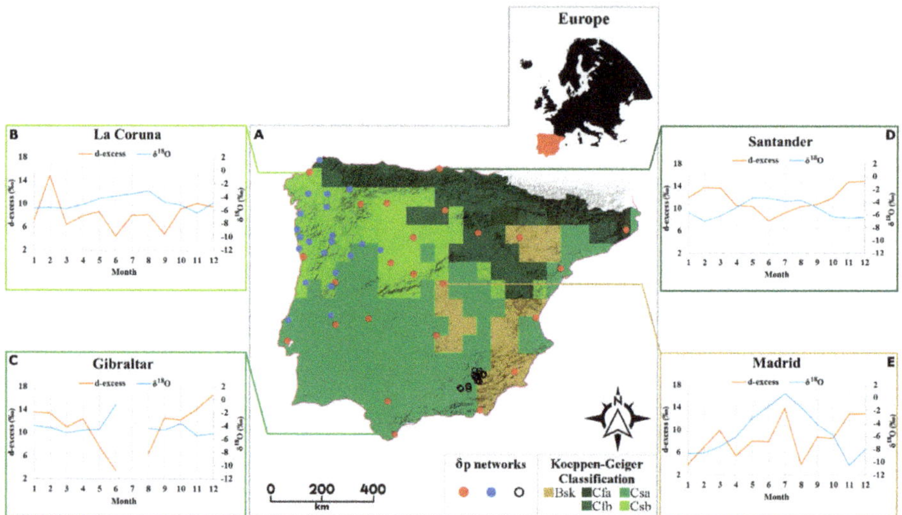

Figure 1. The Iberian Peninsula, with precipitation stable isotope measurement networks and Köppen climate zones [34]. The 24 stations of the Global Network of Isotopes in Precipitation (GNIP) network [23] functioning between October 2002–October 2007 (red solid circles), and two regional networks (NW: [35]; and SE: [36]) are marked with blue solid circles and black circles, respectively (**A**). The seasonal cycle of $\delta^{18}O_p$ and d-excess of precipitation of the four characteristic climate zones (**B**, **C**, **D**, **E**) was computed for mean monthly data of a common six-yr-long (2001–2006) reference period. The color coded Köppen climate zones [34] overlaid on the World_Terrain_Base basemap (Sources: Esri, USGS, NOAA) in ArcMap 10.5 [37]. The red area on the Europe inset map shows the Iberian Peninsula.

The opportunity arose to expand the GNIP dataset with precipitation stable isotope data from two regional monitoring campaigns (Figures 1A and 2). In the first case, 21 stations were available in NW Iberia (October 2004–June 2006) [35]. In the second case, 10 stations (May 2006) and 19 stations (April 2007) were available from SE Iberia [36]. The SE regional network is located in a spatially more confined mountainous area with abrupt differences in elevation between neighboring sites, while that in the NW is better distributed, both vertically and horizontally over a larger region. Unfortunately, merging the GNIP and SE Iberia datasets drew attention to a critical lack of spatial homogeneity (Figures S2–S4), and this turned out to be unsuitable for use in the geostatistical analysis (for details, see Electronic Supplementary Materials), unlike the combination of the GNIP and the NW network. In the latter case, the average number of stations with available records from the GNIP database (~13) increased to ~34 for the period October 2004 to June 2006 (Figures 1 and 2).

The precipitation records archived alongside monthly δ_p values in the GNIP database were found in some cases to have gaps and/or be erroneous, raising questions regarding the calculation of seasonal and annual precipitation-amount weighted averages from monthly δ_p values. Therefore, gridded (0.5° × 0.5°) monthly rainfall totals from the Global Precipitation Climatology Centre (GPCC) database [38], derived as precipitation anomalies at stations interpolated and then superimposed on the GPCC Climatology V2011 [39], were utilized instead. Nevertheless, there were some discrepancies between the monthly precipitation in the GNIP records and the corresponding GPCC values: $r_{min} = 0.82$ (Gibraltar), $r_{max} = 0.99$ (Almeria Aeropuerto), and $r_{mean} = 0.95$ (17 GNIP stations) for 2000–2013. An obvious advantage of the gridded data set is the lack of gaps and the fact that it is less prone to single-station error.

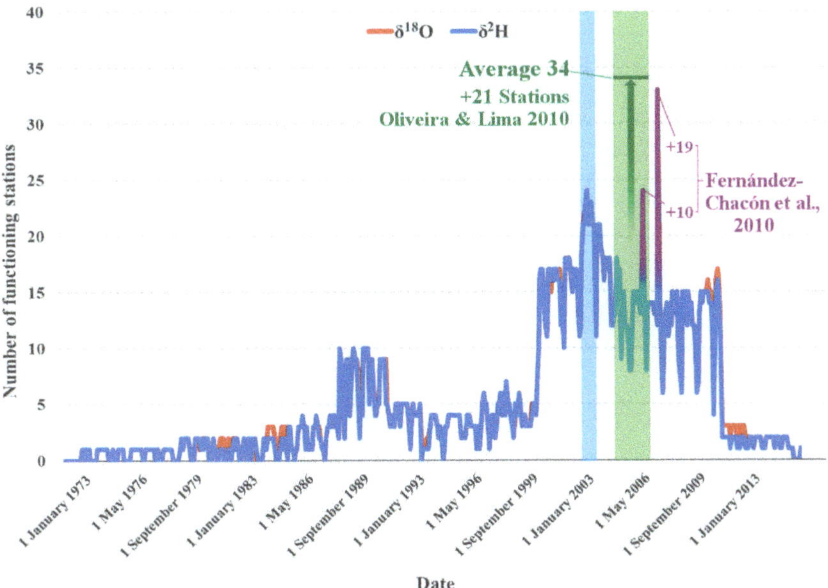

Figure 2. Number of δ_p records obtained from the GNIP stations and the NW [35] and SE regional networks [36] for the period 1973–2015, during which continuous measurements were performed. The blue vertical rectangle represents the period for which only the GNIP δ_p records were assessed (October 2002–September 2003; avg. 21 stations). The green rectangle marks the period when the GNIP data was merged with the δ_p records from the NW network (October 2004–June 2006). The horizontal green line represents the average number of functioning stations [35] in the latter period. The purple lines indicate the number of records of the merged GNIP and SE network [36] for May 2006 and Apr 2007, resulting in 24 and 33 records, respectively.

2.3. Methodology

2.3.1. Steps of the Analysis and Preprocessing

The first step in data preprocessing was to numerically check the δ_p values for database errors (e.g., typos, sign errors) [40]. For example, very negative d-excess values, occasionally even lower than −10‰, corresponding to a monthly precipitation total exceeding 5 mm was all carefully compared to the surrounding stations. The negative d-excess value was interpreted as evidence of evaporative enrichment of the sample and it was discarded from the evaluation if no neighboring stations reported similarly extreme d-excess value for the given month. In addition, in July 2003 at GNIP station Madrid, somewhat improbably, the same values were found as in June, without any precipitation in July being recorded at that site. Consequently, the fact that the June values reappeared in July was suspected to be a database error, and was thus omitted.

Due to the low amount, or frequently the complete lack of precipitation in summer in southern and eastern Iberia (Figure 1), the May to August monthly data were discarded from the analysis (grey bars in Figure 3). Nevertheless, the summer (defined as April to September) seasonal average was calculated, despite the unrealistically low values of d-excess and high values of δ_p, because the corresponding precipitation amounts were also low, having only a small impact on the amount weighted summer seasonal value.

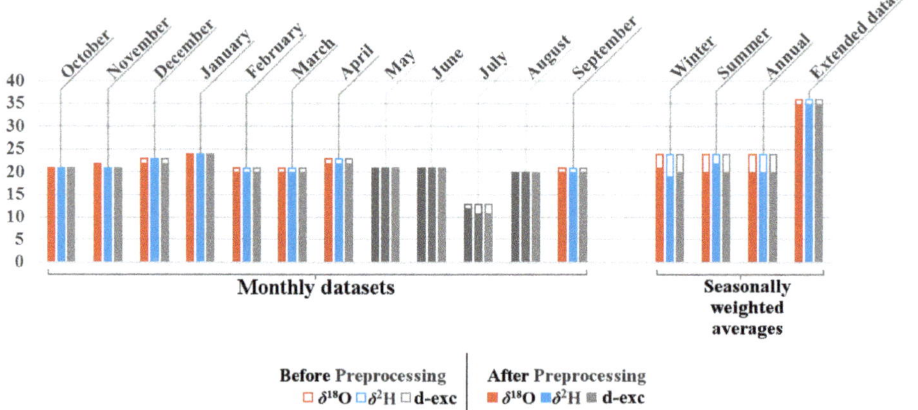

Figure 3. The number of precipitation stable isotope records in the initial monthly and multi-monthly datasets before and after preprocessing entered into the semivariogram analysis. Winter: October–March; summer: April–September. Grey bars represent the months omitted from the analysis, see text for details. The open part of a column indicates that some values were discarded after the quality check of the raw data (see Section 2.3.2).

An amount-weighted mean was only calculated for a period (season, year) if at least 75% of the fallen precipitation was analyzed for the given isotope. This required completeness is a bit stricter criterion than the GNIP protocol (70%) [40].

2.3.2. Multiple Regression Analysis

Trend-like tendencies in δ_p over large distances across the Iberian Peninsula have been documented [24,29,35]. These empirical relationships of δ_p with geographical factors (e.g., altitude effect) may mask the finer scale spatial autocorrelation patterns, which are the focus of interest in the present investigation. The effect of the geographical factors influencing the water stable isotopes' variability in precipitation first has to be minimized in order to obtain representative results with variography (Section 2.3.3), i.e. the spatial trend has to be removed from the data to obtain second-order stationarity [41]. Thus, the effect that was linked to the geographical factors was minimized by computing best-fit multiple regression models and determining their residuals.

The investigation was conducted while using the stepwise procedure suggested by O'Brien, with latitude, longitude (Web Mercator EPSG: 3857), elevation, and geodetic distance from the coast as the independent predictor variables, and the primary precipitation stable isotope values as the dependent variables [42]. Of the four independent variables, the first three were found to be applicable to the minimalization of the effect of geographical factors influencing the raw δ_p records. Probabilities ranged between $p < 2.9 \times 10^{-8}$ and $p < 0.04$, variance inflation was negligible (Variance Inflation Factor < 1.187), and the average adjusted values of R^2 were 0.63 and 0.58 for $\delta^{18}O_p$ and δ^2H_p, respectively (Table S1). No trend removal was found to be necessary in the case of the derived isotopic parameter (d-excess). A similar multiple regression approach was applied to determine the degree of dependence of δ_p on the geographical variables in Central America [43]. However, the advantage of the present approach is that it uses Variance Inflation Factor to control possible multicollinearity [42].

Moran's I global statistic [44] was calculated (binary weighting, bandwidth = 4) for all three isotopic parameters, to investigate whether the residuals carry significant spatial information vital for the derivation of the desired isoscapes.

2.3.3. Variography

The basic function of geostatistics, the variogram, was used to explore the spatial autocorrelation structure of δ_p and d-excess in the Iberian Peninsula. The empirical semivariogram might be calculated while using the Matheron algorithm [45], where $\gamma(h)$ is the semivariogram and $Z(x)$ and $Z(x+h)$ are the values of a parameter sampled at a planar distance $|h|$ from each other

$$\gamma(h) = \frac{1}{2N(h)} \sum_{i=1}^{N(h)} [Z(x_i) - Z(x_i + h)]^2 \tag{1}$$

$N(h)$ is the number of lag-h differences. The most important properties of the semivariogram are the nugget, quantifying the variance at the sampling location (including information regarding the error of the sampling), the sill (c), that is, the level at which the variogram stabilizes, and the range (a), which is the distance within which the samples have an influence on each other and beyond which they are uncorrelated [46]. Note here, that reported ranges in the study are planar distances (d_{planar}) in km (EPSG: 3857), unless otherwise reported; estimated average conversion in the region: d_{planar} × 0.755 = $d_{geodetic}$. In practice, an obtained range is usually attributed to environmental processes that act on the same scale as the range of the specific variable [46], and they are known to have a physical relationship with the assessed variable. In addition, a nugget-type variogram is obtained if the semivariogram does not have a rising part and the points of the empirical semivariogram align parallel to the abscissa. In this case, the sampling frequency is insufficient for estimating the sampling range while using variography [13].

The semivariograms that were derived from the monthly GNIP data required 10 bins and a maximum lag distance of 540 km to achieve the most uniform number of station pairs in the analysis, while, in the case of those from the extended dataset, these figures were 20 bins and a maximum lag distance of 660 km. For geostatistical modeling (e.g., kriging), theoretical semivariograms have to be used to approximate the empirical ones [47]. In practice, the spherical and Gaussian models are among the most commonly used theoretical semivariograms. These were, in fact, applied in the present study by least squares model fitting [48]. While a spherical model displays a linear behavior at the origin and the sill is reached at a, a Gaussian has a repressed slope (tangent with zero slope at the origin), and its sill is a limited value that might be approximated to an infinitesimal degree, but never reached. Thus, in the case of the Gaussian model, a practical and an effective range $a_e = a × \sqrt{3}$ have to be determined; a_e is then used for further geostatistical modeling [46,49].

As the final step in the data preprocessing, semivariogram clouds were utilized to search for additional outlying values. These then had to be omitted following a detailed investigation, resulting in the final set and number of data as the input for the variography (Figure 2). The amount weighted seasonal averages were only calculated after this step has been taken; hence, monthly values that were omitted on the basis of the semivariogram clouds were also treated as missing isotopic parameters.

2.3.4. Isoscape Derivation with Kriging

The isoscape was derived with a three step residual kriging procedure [12,13,31,50]. As a first step, an *initial grid* (10 × 10 km) of stable isotope variation was derived in the region described by the multiple regression model of the supposedly driving geographic variables (LAT, LON, ELE; for details, see Sect. 2.3.2). Next, an interpolated (ordinary point kriging) *residual grid* was modelled while using the theoretical semivariogram (Section 2.3.3) that was fitted on to the residuals of the multivariate regression model. Note here that a methodological comparison testing various interpolation methods for the residual field of δ_p in the central and eastern Mediterranean region reported that the best results were obtained while using ordinary kriging to derive the *residual grid* [50]. Finally, the *initial* and *residual* grids were summarized to obtain the final isoscape.

All of the computations were performed while using Golden Software Surfer 15, GS+ 10, and the raster [51] and lctools [52] packages in R [53]. The digital elevation model was retrieved from Shuttle

Radar Topography Mission data [54] while using Global Mapper. Gimp 2.8 and MS Excel 365 were used for certain visualizations of the results.

3. Results and Discussion

The obtained significant ($p < 0.05$) global Moran' Is for each isotopic parameter range between 0.18 and 0.27, indicating positive spatial autocorrelation in the residuals. This encouraged the incorporation of the 'residual spatial patterns' in the derivation of the isoscapes with variography.

3.1. Variography Using GNIP Data

Variogram analysis was first conducted on monthly residuals of the δ_p records of GNIP stations, and then the amount weighted seasonal averages (winter: October–March; summer: April–September) and the annual residuals (Table S2). With regard to $\delta^{18}O_p$, three of 12 the semivariograms (two monthly, January, March see Figure 4A, and one seasonal, summer) were of the nugget-type, while, in the case of δ^2H_p, this was true for two of the 12 cases (two monthly: January, March). Turning to d-excess, eight out of 12 semivariograms (five monthly: October, November, January, February, March see Figure 4B, 2 seasonal: summer, winter, and the annual) were of the nugget-type.

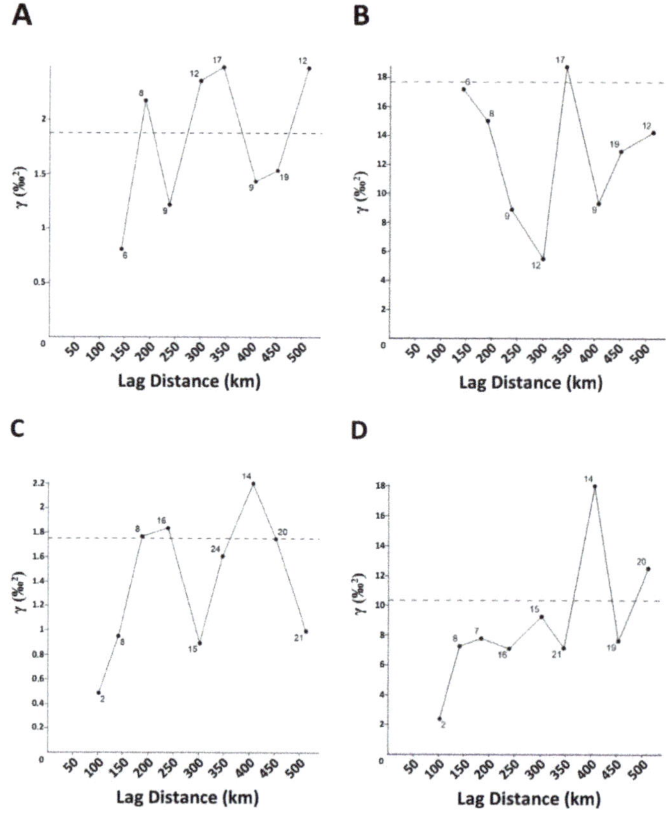

Figure 4. Empirical semivariograms of $\delta^{18}O_p$ and d-excess for March 2003 (**A**, **B**, respectively) and December 2002 (**C**, **D**, respectively). The semivariograms from the March data are of the nugget-type, while the December ones have a rising section. The numbers next to the empirical semivariogram (black dots) show the data pairs within a given distance (bin width was 54 km) behind the semivariograms. The dashed line indicates the average variance.

Unfortunately, a fair portion of the theoretical semivariograms proved to be unsatisfactory due to the lack of data from short distances. There is no month for which it can be said that there are both (i) stations sampling within a 50 km of each other and (ii) more than two stations sampling simultaneously in a 100 km distance. Worse still, in e.g., March 2003, there was no station pair sampling within a 100 km radius (Figure 4A,B), a critically low degree of coverage. Therefore, the small variances over short distances were not observed, only large variance over long distances. Unfortunately, the fitting of theoretical semivariograms was not performed because of the low number and the lack of station pairs at the shortest distances, even in those cases where the semivariogram did outline a rising section (Figure 4C,D). Besides the spatial deficiency of the GNIP network discussed above, the partial evaporation of raindrops [55] during the events with only a few millimeters of precipitation—typical in hot Mediterranean summers—can also result in local noise, further encumbering the geostatistical analysis of the δ_p signal. The fact that unusual δ_p characteristics were reported for S Europe [56,57] during the severe European heat waves of 2003 [58] seems to concur with this statement.

3.2. Variography Using the Extended Database

There was a unique opportunity to extend the δ_p records of the GNIP stations with an additional dataset [35] covering a two year period (October 2004–June 2006; Figures 1 and 2). The station network of this regional ad hoc sampling campaign provided a seamless union with the coexisting GNIP data set (Figure S1). The merged dataset had a greater amount of short distance data pairs than that from the GNIP stations. The Gaussian semivariograms that were fitted to the empirical variograms of $\delta^{18}O_p$ (Figure 5A) and δ^2H_p (Figure S5) obtained from the merged GNIP & NW dataset indicated a ~510 km and a ~420 km range, respectively. For the d-excess, a spherical semivariogram indicated a range of ~440 km (Figure 5B). If these ranges are considered to define the isotropic impact areas, the extended network provides full spatial coverage of the peninsula regarding all three parameters (Figure 5 and Figure S5).

In the meanwhile, a shorter spatial range (below ~100 km) seems to be present in the case of the primary isotopic variables, however this remains speculative and needs further investigation.

The determined spatial range (average ~450 km in planar distance) indicated by the primary isotopic parameters and the range of d-excess refer to a substantial change in the spatial correlation structure of isotopes in precipitation at the scale of the zonal extension of the peninsula. The spatiotemporal variability of moisture contribution over the Iberian Peninsula can link this observation. Studies agree that the Atlantic Ocean and the Mediterranean Sea are the primary marine moisture sources, with a varying degree of spatial- and intra-annual dominance over the peninsula [26,59]. In addition, a characteristically different isotope-hydrological signature is observed between the Mediterranean (d-excess ~16‰) and the Atlantic (d-excess ~10‰) moisture sources (e.g., [28,60]). The recently documented division at ~5°W longitude separating the western part of the peninsula with no influence from the Mediterranean moisture in winter, from the eastern part where western Mediterranean moisture is prevailing during spring and summer seasons [27], is well in line with the above interpretation.

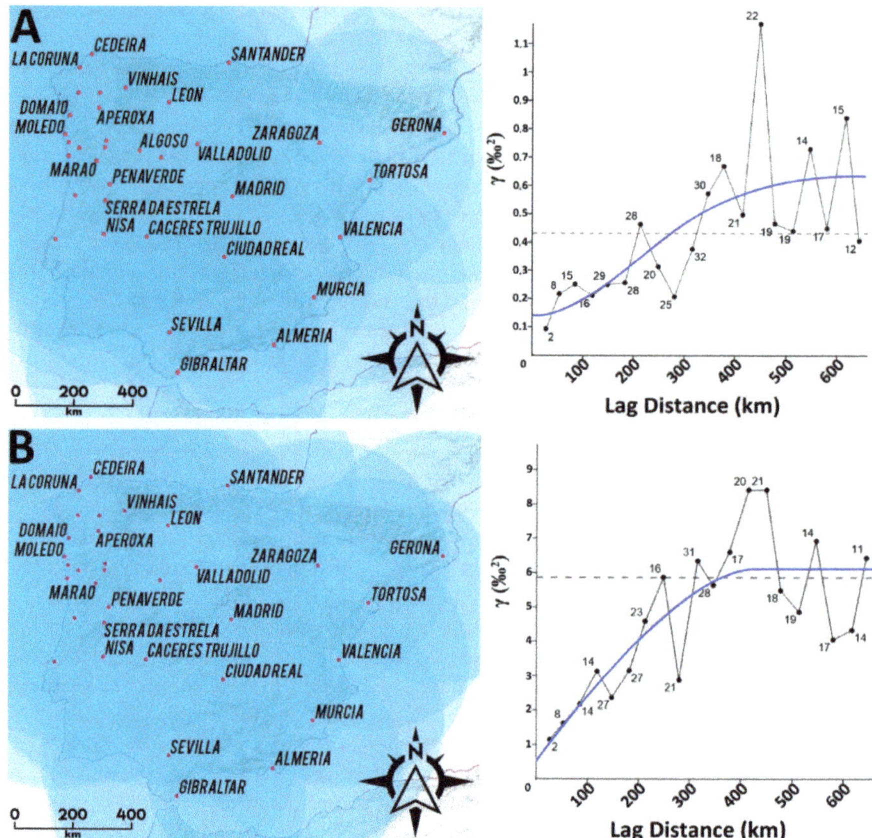

Figure 5. Maps representing the sampling sites (left panels: red dots), the range ellipses around them (blue circles) and the empirical (black line with dots in right panels) and fitted theoretical semivariogram (continuous lines in right panels) derived from $\delta^{18}O_p$ (**A**) and d-excess (**B**). For $\delta^{18}O_p$ (**A**) the parameters of the Gaussian semivariogram model fitted were $C_0 = 0.135$; $C_0 + C = 0.635$; $a_e = 507$ km; $r^2 = 0.49$; RSS = 0.644. For d-excess (**B**) the parameters of the spherical semivariogram were $C_0 = 0.51$; $C_0 + C = 6.105$; $a = 439$ km; $r^2 = 0.66$; RSS = 29.7. The bin width was 33 km; the dashed horizontal line represents the variance. The figure represents the period October 2004–July 2006.

3.3. Isoscape of Precipitation Oxygen and Hydrogen Stable Isotopes for the Iberian Peninsula

The *initial grid* of $\delta^{18}O_p$ derived from the multiple regression model of the "Extended dataset" (Table S1) was combined with the *residual grid* that was obtained with the Gaussian semivariogram while using Equation (2) to derive the estimated values of $\delta^{18}O_p$ (Figure 6A)

$$\widehat{\delta^{18}O_p} = -4.9592 - 1.3634 * 10^{-6} * A - 1.8201 * 10^{-7} * B - 1.6752 * 10^{-3} * C + D \qquad (2)$$

where A and B stand for zonal and meridional Web-Mercator coordinates (EPSG: 3857), respectively, C for elevation and D for the *residual grid*. With a similar procedure, the δ^2H_p isoscape was also derived (Table S1, Figure S6).

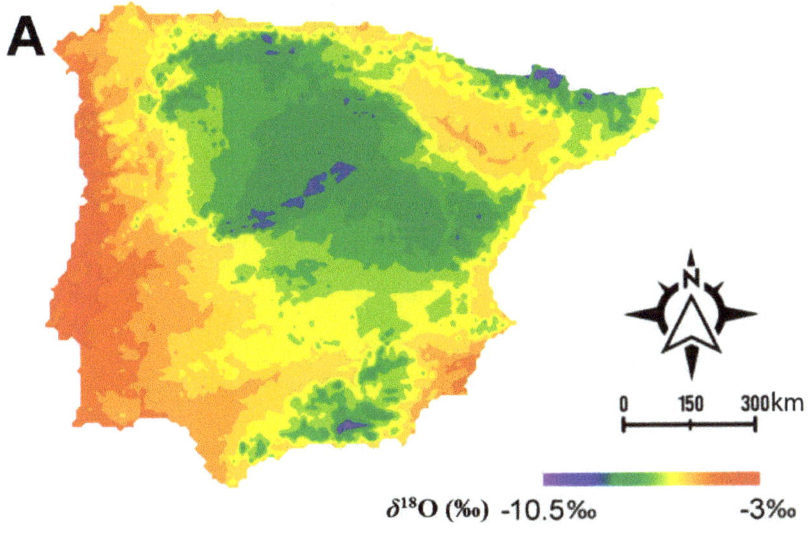

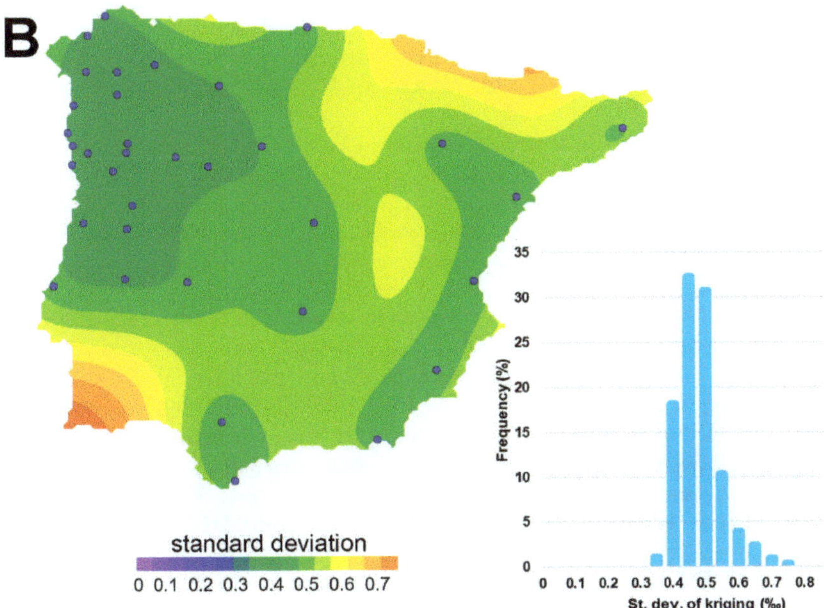

Figure 6. Predicted amount-weighted $\delta^{18}O_p$ isoscape for the Iberian Peninsula for October 2004–July 2006 (**A**) and the standard deviation of kriging (**B**). Grid resolution: 10×10 km. The histogram in panel (**B**) shows the frequency/abundance of the kriging error in 0.1 bins. The north arrow and scale in panel A is relevant for both maps and the black dots represent the precipitation sampling sites used in the geostatistical analysis representing October 2004–July 2006.

The $\delta^{18}O_p$ values that were indicated by the obtained isoscape range between −10.55 and −3.97‰ (Figure 6A). The estimations error is < 0.5‰ at the stations, while the largest error (~0.75‰) is obtained in areas where station data were not available (as usually observed, see [14]) e.g., in the northern (Pyrenees) and SW part of the peninsula (Algarve) (Figure 6B). A similar pattern is seen for δ^2H_p with values ranging from -76.6 to $-21.1‰$ (Figure S6A) and estimated errors up to 6‰ (Figure S6B).

The presented δ_p isoscapes for the Iberian Peninsula can be compared to former products (models and maps) predicting/featuring the spatial distribution of amount-weighted long-term precipitation water stable isotopes across the Iberian Peninsula. A map of $\delta^{18}O_p$ distribution in peninsular Spain and the Balearic Islands was produced using composite monthly samples for the period 2000–2006 of 15 GNIP stations, belonging to the Spanish network for isotopes in precipitation (REVIP) in a polynomial model while employing latitude and elevation as predictors [29]. The REVIP-based map is not available in numerical format, so only the mapped features and some methodological differences can be evaluated. The most striking similarity is that both the current $\delta^{18}O_p$ isoscape (Figure 6A) and the REVIP-based map [29] capture the orography-induced isotopic depletion for high reliefs. The polynomial model with geographical factors can be regarded as an analogue of the initial grid of this study. Thus, the added value of the presented isoscape is the consideration of the spatial variability that is retained in the model residuals irrefutably carrying sub-regional information. For example, zonal heterogeneity of vapor sources [13,14,31], which is especially interesting information from an isotope-hydrometeorological perspective and must be taken into account to improve regional isotopic landscapes. Indeed, ignoring the $\delta^{18}O_p$ residual variance might inspire the debatable conclusion that Atlantic Ocean is the main source of water vapor over the region [29], since the isotope-hydrometeorological dichotomy reflecting the dual moisture sources, clearly also expressed by recent atmospheric transport models [26,27], could be captured in the residual field. Finally, technical advancements in the current $\delta^{18}O_p$ isoscape when compared to the REVIP-based map are that the extended station network (14 more stations from Portugal, 6 from Spain and another one from Gibraltar) definitely (i) improved the accuracy of both the fitting of the regression model and the subsequent interpolation and (ii) extended the coverage of the mapping to the entirety of the peninsula.

Data products and estimated spatial errors of the global regionalized cluster-based water isotope prediction (RCWIP) model [16] are both freely available [61], so detailed numerical comparison between the Iberian subset of RCWIP model and the current regionally fitted δ_p isoscapes could be calculated by subtracting the present model values from the resampled (10×10 km) RCWIP ones. Their difference maps (Figure 7 and Figure S7) indicate more positive values that were obtained by the current model over the mountains: such as the Pyrenees, the Cantabrian Massifs, in a small patch covering the peak region of the Beatic Cordilleras, and along the Iberian System. Major regional patterns in the difference maps are (i) lower values in the RCWIP over extended part of the Atlantic and Cantabrian regions (difference $< -2.5‰$ for $\delta^{18}O_p$ and $< -5‰$ for δ^2H_p), while (ii) positive differences along the southernmost part of the peninsula (difference $> 1.5‰$ for $\delta^{18}O_p$ and $> 10‰$ for δ^2H_p) indicate that RCWIP model predicted higher values south from the Beatic Range (Alboran Coast). Over a remarkably large part of the central (e.g., Meseta, Andalusia) and northeastern part (e.g., Ebro Basin, including the coast of the Balearic Sea) of the peninsula the predicted δ_p values agree (Figure 7 and Figure S7). Two climatic zone domains in the RCWIP model cover the Iberian Peninsula [16], however their boundary does not separate the positive and negative field seen in the difference maps (Figure 7 and Figure S7). Moreover, the predicted values agree very well for extended areas within both RCWIP zones, so the regional scale differences likely do not reflect the RCWIP zones. The differences cannot be explained by the incorporation of the additional samples from the northwestern sector of the peninsula [35], either, because (i) the region of negative difference also extends to the Cantabrian Coast, from where no station was included in the complementary dataset and (ii) the field of positive anomalies along the Alboran Coast is situated >500 km form the area covered by the complementary dataset. It is suggested that the regional patterns seen in the difference map are evidence of a considerable improvement in the current regional prediction of the spatial distribution of long-term mean δ_p across the Iberian Peninsula.

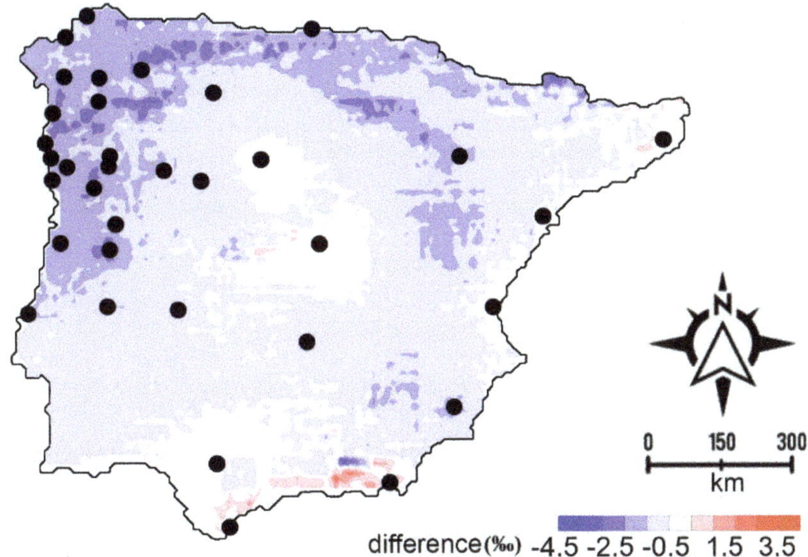

Figure 7. Difference map of the Iberian subset of the regionalized cluster-based water isotope prediction (RCWIP) model and the one presented in the study. For each grid point the values of the Iberian $\delta^{18}O_p$ isoscape (Figure 6A) were subtracted from the RCWIP model. Grid resolution: 10×10 km, black dots represent the precipitation sampling sites used in the geostatistical analysis representing October 2004–July 2006.

The kriging error of the presented Iberian precipitation isoscape (as seen from the in-set histograms; Figure 6B and Figure S6B) is smaller by more than a factor of two for the Iberian Peninsula when compared to the RCWIP model [16], where it ranged between 0.96–0.98‰ for $\delta^{18}O_p$ and 7.9–8.1‰ for δ^2H_p. Thus, the presented isoscape provides a more precise estimation of δ_p than the regional subset of the RCWIP model. This is in harmony with the statement, where the derivation of regional isoscapes is suggested to better estimate the regional variability of δ_p [16], since these do take local geographical features or meteorological parameters into account, which could influence local rainfall patterns leading to a good agreement between observed and modelled data, unlike global isoscapes [11].

It should be noted that these differences in mean predictions might partly reflect different temporal coverage. The RCWIP is a long-term climatological average that is based on pooled non-continuous decadal-scale data for 1960–2009 for δ_p, while the Iberian isoscape covers a narrower interval (2004–2006). Moreover, the differences in prediction uncertainty are almost certainly affected by different time-averaging i.e. the uncertainty in RCWIP model parameters includes uncertainty that is associated with predicting average values at stations that sample different periods of time, while, in the case of the present Iberian isoscape, this was not the case.

4. Conclusions

The monthly GNIP δ_p records (October 2002–September 2003) for 32 stations distributed across the Iberian Peninsula appear to be unsatisfactory for use in the determination of a sampling range of isotope hydrometeorological parameters with variography. However, extended by the addition of a 21-station regional dataset (for the period October 2004–June 2006), a much denser spatial coverage was obtained, proving to be suitable for the exploration of the spatial autocorrelation structure of δ_p across the Iberian Peninsula. Gaussian semivariograms were modelled for δ^2H_p and $\delta^{18}O_p$, and an additional one for d-excess with ranges corresponding to ~450 km in planar distance (~340 km in geodetic distance). The ~450 km spatial range is thought to be related to hydroclimatological processes prevailing on the scale of the zonal extension of the peninsula and they could be connected to a spatiotemporal switch between Atlantic and Mediterranean moisture sources. A sparser network (e.g., GNIP) might also prove to be representative since the obtained spatial range of the δ_p monitoring network provides a coverage for the peninsula. It means that the GNIP monitoring network is suitable for exploration of large-scale isotope hydrological processes/phenomena in the peninsula, but its verification could only be done with a denser network. Finally, the results encourage the development of precipitation stable isotope models at a sub-continental scale in further regions.

Supplementary Materials: The following are available online at http://www.mdpi.com/2073-4441/12/2/481/s1, SuppText.pdf: Additional methodological description (including Figures S1 to S6, and Table S1), Table S2: Stable isotopic data used in the study. Additionally, predicted amount-weighted isoscapes (Figure 6A and Figure S6A) are provided in GEOtiff format.

Author Contributions: Conceived and designed the study: Z.K., I.G.H. Performed the analysis and produced the figures: I.G.H., D.E. Analyzed the data: Z.K., E.D., I.G.H. Wrote the paper: Z.K., P.V., I.G.H., E.D. We applied the FLAE approach for the sequence of authors; see 10.1371/journal.pbio.0050018. All authors have read and agreed to the published version of the manuscript.

Funding: This research was funded by the National Research, Development and Innovation Office; under Grant SNN118205; Slovenian Research Agency ARRS under Grant: P1-0143 and N1-0054; the MTA under Grant "Lendület" program (LP2012-27/2012), and the Ministry of Human Capacities under Grant NTP-NFTÖ-17-B-0028.

Acknowledgments: The authors would like to thank Paul Thatcher for his work on our English version. This is contribution No. 67 of 2ka Palæoclimatology Research Group. This paper is dedicated to the memory of Antal Füst (1940-2020), a pioneer of Geomathematics.

Conflicts of Interest: The authors declare no conflict of interest.

References

1. Dansgaard, W. Stable isotopes in precipitation. *Tellus* **1964**, *16*, 436–468. [CrossRef]
2. Fórizs, I. Isotopes As Natural Tracers In The Watercycle: Examples From The Carpathian Basin. *Studia UBB Phys.* **2003**, *1*, 69–77.
3. Bowen, G.J.; Good, S.P. Incorporating water isoscapes in hydrological and water resource investigations. *Wiley Interdiscip. Rev. Water* **2015**, *2*, 107–119. [CrossRef]
4. Araguas-Araguas, L.; Danesi, P.; Froehlich, K.; Rozanski, K. Global monitoring of the isotopic composition of precipitation. *J. Radioanal. Nucl. Chem.* **1996**, *205*, 189–200. [CrossRef]
5. Rozanski, K.; Araguás-Araguás, L.; Gonfiantini, R. Isotopic patterns in modern global precipitation. In *Climate Change in Continental Isotopic Records*; Swart, P.K., Lohmann, K.C., McKenzie, J., Savin, S., Eds.; American Geophysical Union: Washington, DC, USA, 1993; pp. 1–36.
6. Yoshimura, K. Stable Water Isotopes in Climatology, Meteorology, and Hydrology: A Review. *J. Meteorol. Soc. Jpn. Ser. II* **2015**, *93*, 513–533. [CrossRef]
7. Liu, Z.; Bowen, G.J.; Welker, J.M.; Yoshimura, K. Winter precipitation isotope slopes of the contiguous USA and their relationship to the Pacific/North American (PNA) pattern. *Clim Dyn* **2013**, *41*, 403–420. [CrossRef]
8. Gibson, J.J.; Edwards, T.W.D.; Birks, S.J.; St Amour, N.A.; Buhay, W.M.; McEachern, P.; Wolfe, B.B.; Peters, D.L. Progress in isotope tracer hydrology in Canada. *Hydrol. Process.* **2005**, *19*, 303–327. [CrossRef]
9. Kong, Y.; Wang, K.; Li, J.; Pang, Z. Stable Isotopes of Precipitation in China: A Consideration of Moisture Sources. *Water* **2019**, *11*, 1239. [CrossRef]
10. Vachon, R.W.; Welker, J.M.; White, J.W.C.; Vaughn, B.H. Monthly precipitation isoscapes (δ^{18}O) of the United States: Connections with surface temperatures, moisture source conditions, and air mass trajectories. *J. Geophys. Res. Atmos.* **2010**, *115*. [CrossRef]
11. Kaseke, K.F.; Wang, L.; Wanke, H.; Turewicz, V.; Koeniger, P. An Analysis of Precipitation Isotope Distributions across Namibia Using Historical Data. *PLoS ONE* **2016**, *11*, e0154598. [CrossRef]
12. Kern, Z.; Kohán, B.; Leuenberger, M. Precipitation isoscape of high reliefs: Interpolation scheme designed and tested for monthly resolved precipitation oxygen isotope records of an Alpine domain. *Atmos. Chem. Phys.* **2014**, *14*, 1897–1907. [CrossRef]

13. Hatvani, I.G.; Leuenberger, M.; Kohán, B.; Kern, Z. Geostatistical analysis and isoscape of ice core derived water stable isotope records in an Antarctic macro region. *Polar Sci.* **2017**, *13*, 23–32. [CrossRef]

14. Lykoudis, S.P.; Argiriou, A.A. Gridded data set of the stable isotopic composition of precipitation over the eastern and central Mediterranean. *J. Geophys. Res. Atmos.* **2007**, *112*. [CrossRef]

15. Chan, W.-P.; Yuan, H.-W.; Huang, C.-Y.; Wang, C.-H.; Lin, S.-D.; Lo, Y.-C.; Huang, B.-W.; Hatch, K.A.; Shiu, H.-J.; You, C.-F.; et al. Regional Scale High Resolution δ18O Prediction in Precipitation Using MODIS EVI. *PLoS ONE* **2012**, *7*, e45496. [CrossRef] [PubMed]

16. Terzer, S.; Wassenaar, L.I.; Araguás-Araguás, L.J.; Aggarwal, P.K. Global isoscapes for $\delta^{18}O$ and δ^2H in precipitation: Improved prediction using regionalized climatic regression models. *Hydrol. Earth Syst. Sci.* **2013**, *17*, 4713–4728. [CrossRef]

17. van der Veer, G.; Voerkelius, S.; Lorentz, G.; Heiss, G.; Hoogewerff, J.A. Spatial interpolation of the deuterium and oxygen-18 composition of global precipitation using temperature as ancillary variable. *J. Geochem. Explor.* **2009**, *101*, 175–184. [CrossRef]

18. Clark, I.D.; Fritz, P. *Environmental Isotopes in Hydrogeology*; Taylor & Francis: Boca Raton, FL, USA; New York, NY, USA, 1997.

19. Hobson, K.A. Tracing origins and migration of wildlife using stable isotopes: A review. *Oecologia* **1999**, *120*, 314–326. [CrossRef]

20. Heaton, K.; Kelly, S.D.; Hoogewerff, J.; Woolfe, M. Verifying the geographical origin of beef: The application of multi-element isotope and trace element analysis. *Food Chem.* **2008**, *107*, 506–515. [CrossRef]

21. Ehleringer, J.R.; Bowen, G.J.; Chesson, L.A.; West, A.G.; Podlesak, D.W.; Cerling, T.E. Hydrogen and oxygen isotope ratios in human hair are related to geography. *Proc. Natl. Acad. Sci. USA* **2008**, *105*, 2788–2793. [CrossRef]

22. Bowen, G.J. Isoscapes: Spatial Pattern in Isotopic Biogeochemistry. *Annu. Rev. Earth Planet. Sci.* **2010**, *38*, 161–187. [CrossRef]

23. International Atomic Energy Agency (IAEA). Global Network of Isotopes in Precipitation. The GNIP Database. Available online: http://www.Isohis.Iaea.Org (accessed on 12 December 2015).

24. Araguas-Araguas, L.J.; Diaz Teijeiro, M.F. *Isotope Composition of Precipitation and Water Vapour in the Iberian Peninsula*; International Atomic Energy Agency: Vienna, Austri, 2005; pp. 173–190.

25. Carreira, P.M.M.; Araujo, M.F.; Nunes, D. *Isotopic Composition of Rain and Water Vapour Samples from Lisbon Region: Characterization of Monthly and Daily Events*; International Atomic Energy Agency: Vienna, Austria, 2005; pp. 141–155.

26. Gimeno, L.; Nieto, R.; Trigo, R.M.; Vicente-Serrano, S.M.; López-Moreno, J.I. Where Does the Iberian Peninsula Moisture Come From? An Answer Based on a Lagrangian Approach. *J. Hydrometeorol.* **2010**, *11*, 421–436. [CrossRef]

27. Ciric, D.; Nieto, R.; Losada, L.; Drumond, A.; Gimeno, L. The Mediterranean Moisture Contribution to Climatological and Extreme Monthly Continental Precipitation. *Water* **2018**, *10*, 519. [CrossRef]

28. Julian, J.C.-S.; Araguas, L.; Rozanski, K.; Benavente, J.; Cardenal, J.; Hidalgo, M.C.; Garcia-Lopez, S.; Martinez-Garrido, J.C.; Moral, F.; Olias, M. Sources of precipitation over South-Eastern Spain and groundwater recharge. An isotopic study. *Tellus B* **1992**, *44*, 226–236. [CrossRef]

29. Rodriguez-Arévalo, J.; Diaz-Teijeiro, M.F.; Castano, S. Modelling And Mapping Oxygen-18 Isotope Composition Of Precipitation In Spain For Hydrologic And Climatic Applications. In *Isotopes in Hydrology, Marine Ecosystems and Climate Change Studies*; Internaional Atomic Energy Agency: Vienna, Austria, 2013; Volume 1, pp. 171–177.

30. Rozanski, K.; Sonntag, C.; Münnich, K.O. Factors controlling stable isotope composition of European precipitation. *Tellus* **1982**, *34*, 142–150. [CrossRef]

31. Bowen, G.J.; Wilkinson, B. Spatial distribution of $\delta^{18}O$ in meteoric precipitation. *Geology* **2002**, *30*, 315–318. [CrossRef]

32. Ninyerola, M.; Pons, X.; Roure, J.M. *Atlas Climático Digital de la Península Ibérica, Metodología Aplicaciones en Bioclimatología y Geobotánica*; Universitat Autónoma de Barcelona, Department de Biologia Animal, Biologia Vegetal i Ecologia (Unitat de Botánica), Department de Geografia: Barcelona, Spain, 2005.

33. Coplen, T.B. Reporting of stable hydrogen, carbon and oxygen isotopic abundances. *Pure Appl. Chem.* **1994**, *66*, 273–276. [CrossRef]

34. Kottek, M.; Grieser, J.; Beck, C.; Rudolf, B.; Rubel, F. World Map of the Köppen-Geiger climate classification updated. *Meteorol. Z.* **2006**, *15*, 259–263. [CrossRef]

35. Oliveira, A.; Lima, A. Spatial variability in the stable isotopes of modern precipitation in the northwest of Iberia. *Isot. Environ. Health Stud.* **2010**, *46*, 13–26. [CrossRef]

36. Fernández-Chacón, F.; Benavente, J.; Rubio-Campos, J.C.; Kohfahl, C.; Jiménez, J.; Meyer, H.; Hubberten, H.; Pekdeger, A. Isotopic composition ($\delta^{18}O$ and δD) of precipitation and groundwater in a semi-arid, mountainous area (Guadiana Menor basin, Southeast Spain). *Hydrol. Process.* **2010**, *24*, 1343–1356. [CrossRef]

37. ESRI. *ArcGIS Desktop, 10.5*; Environmental Systems Research Institute: Redlands, CA, USA, 2016.

38. Becker, A.; Finger, P.; Meyer-Christoffer, A.; Rudolf, B.; Schamm, K.; Schneider, U.; Ziese, M. A description of the global land-surface precipitation data products of the Global Precipitation Climatology Centre with sample applications including centennial (trend) analysis from 1901–present. *Earth Syst. Sci. Data* **2013**, *5*, 71–99. [CrossRef]

39. Meyer-Christoffer, A.; Becker, A.; Finger, P.; Rudolf, B.; Schneider, U.; Ziese, M. *GPCC Climatology Version 2011 at 0.5: Monthly Land-Surface Precipitation Climatology for Every Month and the Total Year from Rain-Gauges built on GTS-based and Historic Data*; Global Precipitation Climatology Centre at Deutscher Wetterdienst: Offenbach, Germany, 2011.

40. International Atomic Energy Agency (IAEA). *Statistical Treatment of Data on Environmental Isotopes in Precipitation*; International Atomic Energy Agency: Vienna, Austria, 1992; p. 781.

41. Hohn, M.E. *Geostatistics and Petroleum Geology*, 2nd ed.; Springer Science+Business Media: Dordrecht, The Netherlands, 1999.

42. O'Brien, R.M. A Caution Regarding Rules of Thumb for Variance Inflation Factors. *Qual. Quant.* **2007**, *41*, 673–690. [CrossRef]

43. Lachniet, M.S.; Patterson, W.P. Use of correlation and stepwise regression to evaluate physical controls on the stable isotope values of Panamanian rain and surface waters. *J. Hydrol.* **2006**, *324*, 115–140. [CrossRef]

44. Moran, P.A.P. Notes on Continuous Stochastic Phenomena. *Biometrika* **1950**, *37*, 17–23. [CrossRef] [PubMed]

45. Matheron, G. *Les Variables Régionalisées et Leur Estimation: Une Application de la Théorie des Fonctions Aléatoires Aux Sciences de la Nature*; Masson et Cie Luisant-Chartres, impr; Durand: Paris, France, 1965.

46. Chilès, J.-P.; Delfiner, P. *Geostatistics*; Wiley: New York, NY, USA, 2012.

47. Cressie, N. The origins of kriging. *Math Geol.* **1990**, *22*, 239–252. [CrossRef]

48. Oliver, M.A.; Webster, R. A tutorial guide to geostatistics: Computing and modelling variograms and kriging. *CATENA* **2014**, *113*, 56–69. [CrossRef]

49. Herzfeld, U.C. *Atlas of Antarctica: Topographic Maps from Geostatistical Analysis of Satellite Radar Altimeter Data : With 169 Figures*; Springer: Berlin/Heidelberg, Germany, 2004.

50. Argiriou, A.; Salamalikis, V.; Lykoudis, S. *Development and Evaluation of a Methodology for the Generation of Gridded Isotopic Data Sets*; Isotopes in Hydrology, Marine Ecosystems and Climate Change Studies, Monaco, 2011; IAEA: Monaco, 2013; pp. 161–169.

51. Hijmans, R.J. Raster: Geographic Data Analysis and Modeling, 3.0–7. 2019. Available online: https://cran.r-project.org/web/packages/raster/index.html (accessed on 10 December 2019).

52. Kalogirou, S. Lctools: Local Correlation, Spatial Inequalities, Geographically Weighted Regression and Other Tools, 3.0-7. 2019. Available online: http://lctools.science/lctools/ (accessed on 10 December 2019).

53. R Core Team. *R: A Language and Environment for Statistical Computing*; R Foundation for Statistical Computing: Vienna, Austria, 2019.

54. Farr, T.G.; Rosen, P.A.; Caro, E.; Crippen, R.; Duren, R.; Hensley, S.; Kobrick, M.; Paller, M.; Rodriguez, E.; Roth, L.; et al. The shuttle radar topography mission. *Rev. Geophys.* **2007**, *45*. [CrossRef]

55. Stewart, M.K. Stable isotope fractionation due to evaporation and isotopic exchange of falling waterdrops: Applications to atmospheric processes and evaporation of lakes. *J. Geophys. Res.* **1975**, *80*, 1133–1146. [CrossRef]

56. Vreča, P.; Brenčič, M.; Leis, A. Comparison of monthly and daily isotopic composition of precipitation in the coastal area of Slovenia. *Isot. Environ. Health Stud.* **2007**, *43*, 307–321. [CrossRef]

57. Longinelli, A.; Anglesio, E.; Flora, O.; Iacumin, P.; Selmo, E. Isotopic composition of precipitation in Northern Italy: Reverse effect of anomalous climatic events. *J. Hydrol.* **2006**, *329*, 471–476. [CrossRef]

58. Schär, C.; Vidale, P.L.; Lüthi, D.; Frei, C.; Häberli, C.; Liniger, M.A.; Appenzeller, C. The role of increasing temperature variability in European summer heatwaves. *Nature* **2004**, *427*, 332. [CrossRef]
59. Krklec, K.; Domínguez-Villar, D. Quantification of the impact of moisture source regions on the oxygen isotope composition of precipitation over Eagle Cave, central Spain. *Geochim. Cosmochim. Acta* **2014**, *134*, 39–54. [CrossRef]
60. Celle-Jeanton, H.; Travi, Y.; Blavoux, B. Isotopic typology of the precipitation in the Western Mediterranean Region at three different time scales. *Geophys. Res. Lett.* **2001**, *28*, 1215–1218. [CrossRef]
61. International Atomic Energy Agency (IAEA). *RCWIP (Regionalized Cluster-Based Water Isotope Prediction) Model—Gridded Precipitation δ18O|δ2H| δ18O and δ2H Isoscape Data*, 1st ed.; International Atomic Energy Agency: Vienna, Austria, 2013.

Article

Regional Isotopic Signatures of Groundwater in Croatia

Željka Brkić [1,*], Mladen Kuhta [1], Tamara Hunjak [2] and Ozren Larva [1]

[1]   Croatian Geological Survey, Department of Hydrogeology and Engineering Geology, 10000 Zagreb, Croatia; mladen.kuhta@hgi-cgs.hr (M.K.); ozren.larva@hgi-cgs.hr (O.L.)

[2]   Kemolab Ltd., 10000 Zagreb, Croatia; tamarahunjak@gmail.com

*   Correspondence: zeljka.brkic@hgi-cgs.hr; Tel.: +385-1-6160-726

Received: 8 May 2020; Accepted: 10 July 2020; Published: 13 July 2020

Abstract: Tracer methods are useful for investigating groundwater travel times and recharge rates and analysing impacts on groundwater quality. The most frequently used tracers are stable isotopes and tritium. Stable isotopes of oxygen ($\delta^{18}O$) and hydrogen (δ^2H) are mainly used as indicators of the recharge condition. Tritium (^3H) is used to estimate an approximate mean groundwater age. This paper presents the results of an analysis of stable isotope data and tritium activity in Croatian groundwater samples that were collected between 1997 and 2014 at approximately 100 sites. The composition of the stable isotopes of groundwater in Croatia originates from recent precipitation and is described using two regional groundwater lines. One of them is applied to groundwater accumulated in the aquifers in the Pannonian part of Croatia and the other is for groundwater accumulated in the Dinaric karst of Croatia. The isotope content shows that the studied groundwater is mainly modern water. A mix of sub-modern and modern water is mostly accumulated in semi-confined porous aquifers in northern Croatia, deep carbonate aquifers, and (sub)thermal springs.

Keywords: stable isotopes; tritium activity; spatial distribution of an approximate groundwater age; karst aquifer system; porous aquifer system; Croatia

---

1. Introduction

Water molecules have isotopic "fingerprints" according to the differing ratios of their oxygen and hydrogen isotopes and their radioisotope activity. The most frequently used isotopes in groundwater hydrology are stable isotopes and tritium, which naturally occur in the environment. Consequently, they are called "environmental isotopes". All three isotopes enter the hydrological cycle via precipitation. Stable isotopes of oxygen ($\delta^{18}O$) and hydrogen (δ^2H) are mainly used as indicators of the recharge condition. Tritium (^3H) is used to estimate the approximate groundwater age.

The global monitoring of the isotopic composition of monthly precipitation is carried out through the Global Network of Isotopes in Precipitation (GNIP), which was established in 1961 by the International Atomic Energy Agency (IAEA) and the World Meteorological Organization (WMO). The objective of this network is the systematic collection of the isotopic data of precipitation and the determination of temporal and spatial variations in the isotopes within precipitation. Isotopic data include the tritium activity concentration and stable isotopes of hydrogen and oxygen isotopes (δ^2H and $\delta^{18}O$ values), as well as climatological data (mean monthly temperature, monthly precipitation amount, and atmospheric water vapour pressure) [1].

The stable isotopes of hydrogen (^1H, ^2H) and oxygen (^{16}O, ^{18}O) are natural environmental tracers with broad applications in groundwater hydrology [2–8]. They help to reveal the groundwater origin and recharge [9–11], as well as the mean groundwater age in some cases [12,13]. The content of stable isotopes in water is affected by the sources of precipitating air masses, which are linked to climate characteristics such as ambient temperature, precipitation amount, wind speed, and humidity.

Fractionation processes are responsible for the spatial and temporal variations in water isotope composition [14–16], as well as orographic effects [17], distance from the coast, and latitude [18,19].

The isotopic composition of groundwater in both karst and alluvial aquifers in Croatia was previously studied for different purposes. The study of general functioning of the karst hydrogeological system [20] and the delineation of catchment areas of karst springs [21] were performed using an analysis of the distribution of stable isotopes in groundwater. Using stable isotope composition, the recharge areas of karst aquifers were considered [22–25]. A lumped parameter approach was used for the groundwater age estimation [12,26]. Several studies have been carried out where isotope composition of precipitation, surface water, and groundwater were employed for identification of groundwater recharge sources and for improvement of the conceptual model of the porous aquifer systems [27–31].

The isotopic composition of groundwater was studied in the neighbouring countries also. Mezga et al. [32] presented the isotopic composition of sampled groundwater that was monitored over three years and performed a comparison with previous studies regarding the isotopic composition of precipitation, surface water, and groundwater in Slovenia. Nikolov et al. [33] concluded that most of the analysed groundwater in Vojvodina (Serbia) can be characterized as modern waters, recharged mostly from precipitation.

Tritium (^{3}H) is the radioactive isotope of hydrogen with a half-life of 12.32 years [34]. It decays into helium−3 (^{3}He). Tritium is naturally produced in the atmosphere by the reaction of cosmic radiation with nitrogen atoms. In hydrology and oceanography, ^{3}H content is expressed in tritium units (TU). Other disciplines may use the ^{3}H activity concentration in Bq (Bequerel), where 1 TU = 0.118 Bq/L [35,36]. The natural ^{3}H activity concentration in precipitation varies within the range of 5 to 10 TU, depending on the location [9,37]. Prior to atmospheric nuclear bomb testing in the 1950s, the mean ^{3}H activity concentrations were approximately 2 to 8 TU [38]. Due to nuclear bomb testing, the maximum concentrations of ^{3}H in the precipitation in the northern hemisphere reached more than 5000 TU in 1963 [39]. The data over the last twenty years show an almost constant mean annual ^{3}H concentration [40,41]. However, seasonal variations are observable. The lowest ^{3}H concentrations are distributed in the winter months. Duliński et al. [40] suggest that these low ^{3}H activity concentrations likely reflect episodes of specific circulation patterns over Europe, where the fast transport of oceanic water vapour from the Atlantic Ocean to central Europe occurs without significant rainout and moisture exchange with the surface of the continent. The maximum monthly ^{3}H activity concentration is observed in the late spring and summer months [23,40,42–47]. The extremely high ^{3}H activity concentrations occasionally recorded are likely linked to episodic emissions of technogenic tritium on the European continent [48]. The possible sources are nuclear power reactors and the applications of artificial tritium in medicine and the watch industry [40].

Since it is relatively geochemically conservative, tritium is an excellent tracer for the investigation of groundwater flow dynamics as well as giving an approximate groundwater age at a time scale of less than 100 years [2,26,49,50]. Groundwater ^{3}H activity concentrations reflect the ^{3}H activity concentration when the water, which contributes to groundwater recharge, was in the contact with the atmosphere. Since ^{3}H decays to ^{3}He, the application of the ^{3}H/^{3}He method, which measures the relative abundance of ^{3}H and ^{3}He in a groundwater sample [49], allows for a more precise definition of the groundwater age.

This study aims to evaluate the regional isotopic signatures of groundwater in Croatia. The study is focused on comparisons between groundwater stable isotopes and precipitation isotope data. To achieve this, we analyse the existing data on stable isotopes (δ^2H and δ^{18}O) and tritium activity (^{3}H) stored in the database of the Croatian Geological Survey (HGI-CGS), as well as data from published papers. Isotopic data are taken from studies on drinking groundwater. An overview of the sampling sites from different aquifer systems in which water samples were collected between 1997 and 2014 is presented, and the descriptive statistics of isotopic data are shown. We also discuss data distribution and the various correlations among the data.

2. Materials and Methods

2.1. Study Area

Croatia is a small southern European country, comprising an area of only 56,538 km^2. From a hydrogeological and even water management standpoint, the country can be divided into two hydrogeological regions: the Northern or Pannonian region and the Southern or Dinaric karst region (Figure 1). Northern Croatia is situated in a southwestern part of the Pannonian Basin where large lowlands are dominant along the two Danube tributaries—the river Sava and Drava and their major tributaries—and along the river Danube in the east-most part of the region. Quaternary sediments form thick and highly permeable aquifers within these lowlands [51,52]. In between the lowlands, there is a hilly region (in some places mountainous), composed mainly of low permeable deposits ranging in age from the Palaeozoic to Quaternary, while only smaller areas are occupied by permeable carbonate deposits of the Triassic age, which represent an important aquifer. The altitudes of these lowlands vary from approximately 120 m a.s.l. in the west to approximately 80 m a.s.l. in the east. In the mountainous region, the altitudes can reach 1000 m a.s.l.

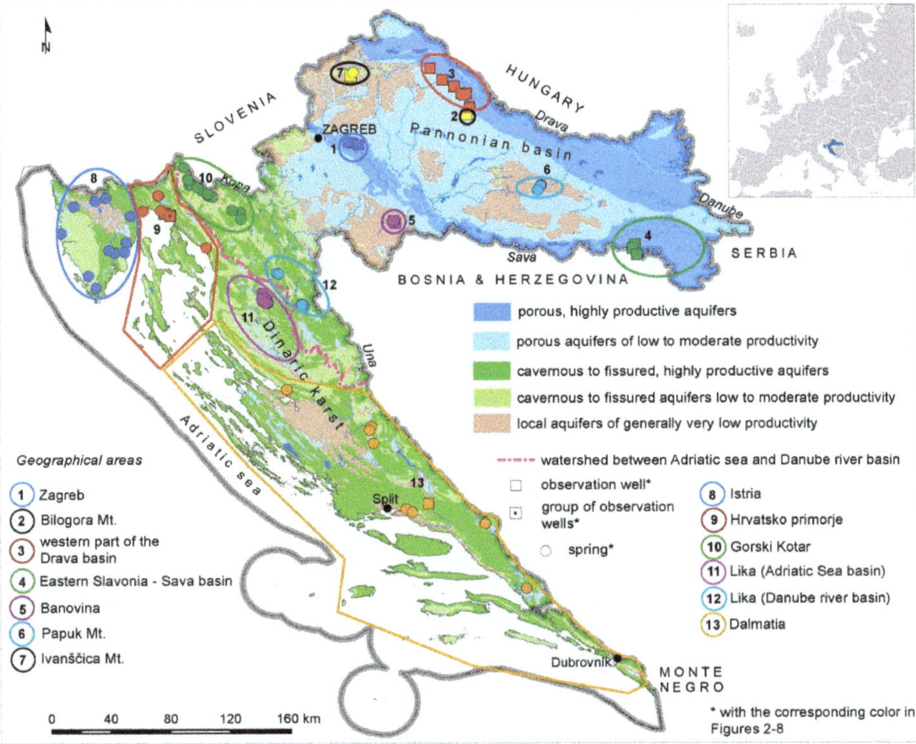

Figure 1. Aquifer types in Croatia, geographical areas (schematic), and the locations of the isotope composition measurements. The colours of the sampling points and the frame colours of the geographical areas correspond to the colours of the points in Figures 2–8.

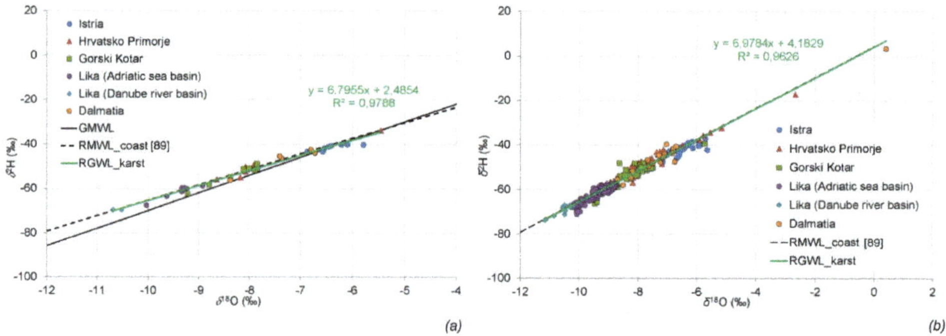

Figure 2. The regional meteoric water line of the coastal part of Croatia (RMWL_coast) and the stable isotope composition of the spring water in the Dinaric karst in Croatia: (**a**) the mean values of the stable isotopes and (**b**) the single values of the stable isotopes.

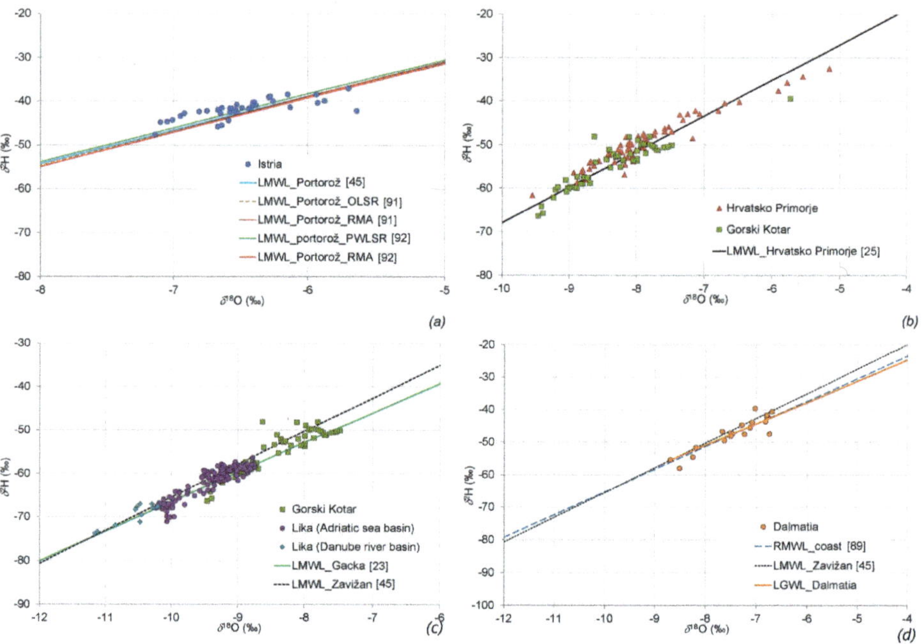

Figure 3. Local meteoric water lines (LMWLs) and the single values of the groundwater stable isotopes in the Dinaric karst in Croatia: (**a**) Istria, (**b**) Hrvatsko Primorje and Gorski Kotar, (**c**) Gorski Kotar and Lika, and (**d**) Dalmatia.

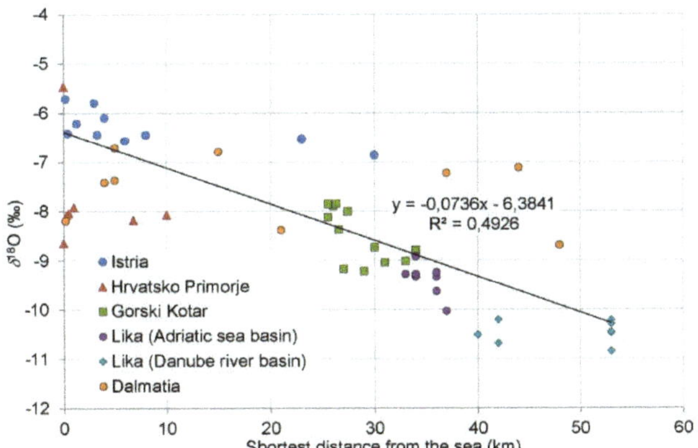

Figure 4. The mean $\delta^{18}O$ vs. the shortest distance from the sea.

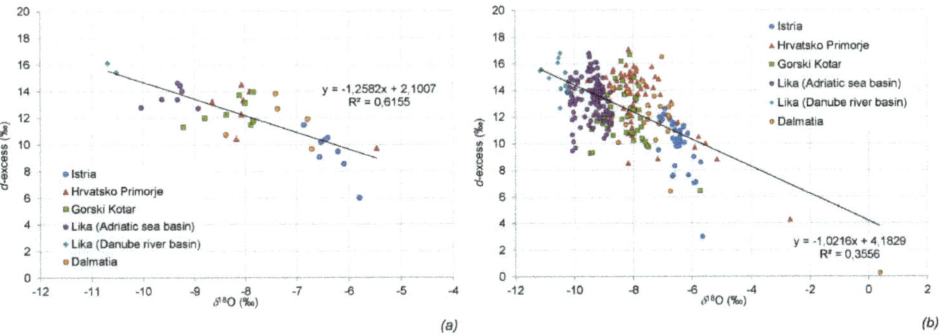

Figure 5. Linear relationship between $\delta^{18}O$ and *d*-excess of karst groundwater: (**a**) the mean values and (**b**) the single values.

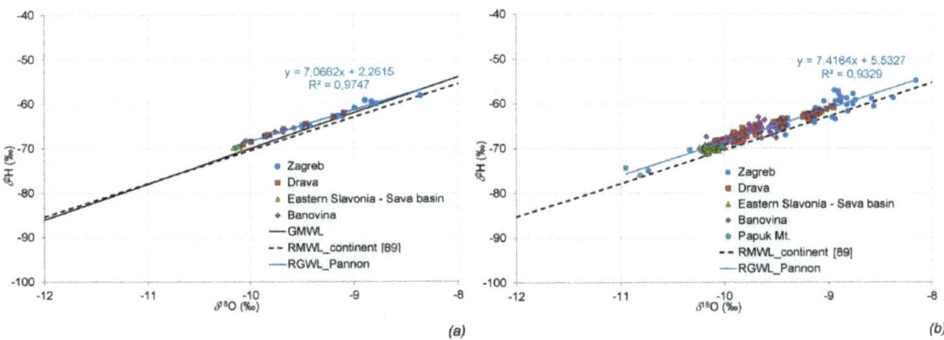

Figure 6. The regional meteoric water line of the continental part of Croatia (RMWL_continent) and the stable isotope composition of spring water in the Pannonian part in Croatia: (**a**) the mean values of stable isotopes and (**b**) the single values of stable isotopes.

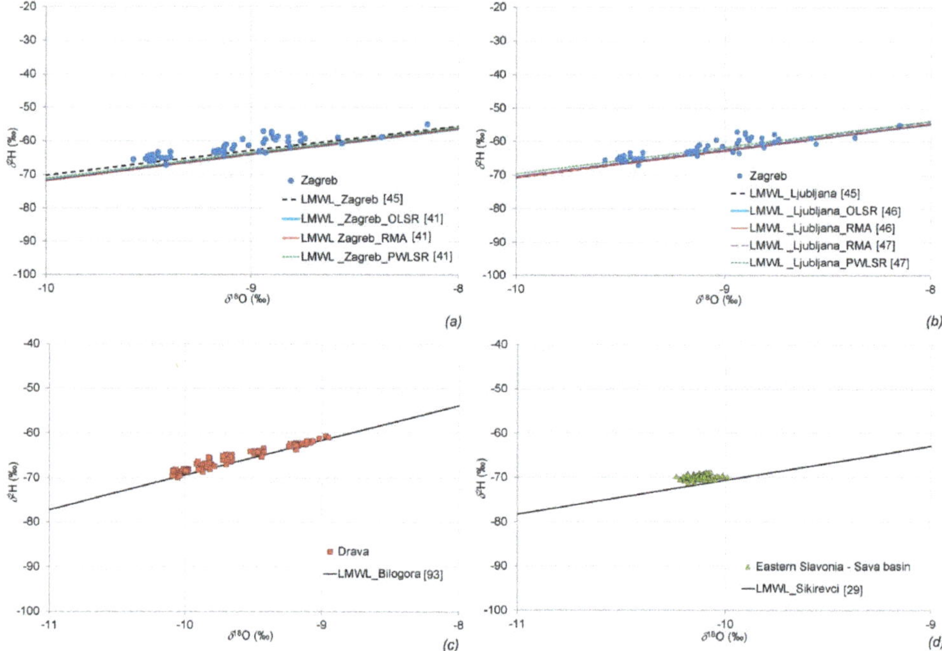

Figure 7. LMWLs and the single stable isotope composition of groundwater in the Pannonian part of Croatia: (**a**) groundwater in the Zagreb aquifer and LMWL Zagreb, (**b**) groundwater in the Zagreb aquifer and LMWL Ljubljana, (**c**) the Drava Basin, and (**d**) Eastern Slavonia—the Sava basin.

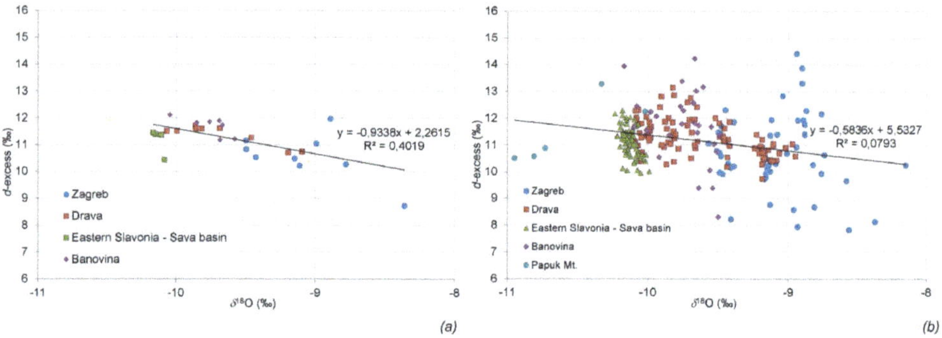

Figure 8. Relationship between $\delta^{18}O$ and *d*-excess in the analysed groundwater in the Pannonian part of Croatia: (**a**) the mean values and (**b**) the single values.

The karst area of Croatia is mainly composed of carbonate rocks of the Mesozoic age. Carbonate rocks, especially limestone, have developed conduit and fissured porosity, which accumulate and transfer a significant amount of groundwater. Besides the existence of numerous geomorphological phenomena, the basic characteristics of the karst area are the absence of surface water, a high velocity of groundwater flow, groundwater discharge at the springs of over ten m^3/s, significant groundwater amounts that discharge into the sea, and the strong impact of the sea on coastal aquifers. There are frequent occurrences here of karst poljes with springs on one side and ponors on the other; this sometimes also means that the same water sinks and reappears multiple times on several levels [poljes] before it reaches the final erosion base [53–55]. The altitudes of these areas vary from approximately 1900 m a.s.l. in the high mountains to 0 m a.s.l. at the coast.

Croatia is divided into two main climate regions—Continental and Mediterranean—with some variations within those climate zones [56]. According to Köppen's classification, most of Croatia has a moderately warm rainy climate (C), whereas only the highest mountain areas have a snow-forest climate (D). The Dalmatian coast and islands have a Mediterranean climate (Cs), while in other parts of Croatia there are different types of moderately warm and humid climates (Cf) which differ according to the warmth of summer months (hot, warm or fresh summers) and annual precipitation. The lowest mean annual air temperature occurs at the highest mountains and reaches 2–3 °C. In the lowlands of northern Croatia, the average air temperature is around 11 °C, while in the Adriatic area, it ranges from 13 °C in the north to 17 °C in the far south. Summer is the driest season along the coast, and winter is the rainiest season, with twice the amount of precipitation as in the summer. The continental part of Croatia has a different climate. The mean annual precipitation in Croatia ranges from about 300 mm to just above 3500 mm. The majority of precipitation is caused by south-western and western air circulation (cyclogenetic area of the Mediterranean), while a significantly smaller amount of precipitation is caused by relatively dry air masses coming from the northeast. The lowest annual precipitation, approximately 300 mm, can be found on the outer islands of the southern Adriatic. On the islands and coasts of central and northern Dalmatia and on the western coast of Istria, about 800 to 900 mm of precipitation can be expected. In continental Croatia, the mean annual precipitation decreases from west to east. In the western part (wider Zagreb area), it ranges from 1000 to 1500 mm, while in Eastern Slavonia, the mean annual precipitation reaches only 600 to 700 mm. Due to orographic influence, the greatest amount of precipitation can be found along the Dinaric Mountain Range, which extends from NW to SE, separating the coastal area from continental Croatia. In this area, the mean annual precipitation reaches 3000 to 3500 mm. The karst aquifers formed below this massif drain towards both regional basins, the Adriatic Sea basin in the south and the Danube basin in the north.

The hydrogeological structures, relief indents, and different climatological influences result in different isotopic signatures of the groundwater in Croatian regions.

2.2. Data and Methods

The first step in analysing the isotopic composition of groundwater was to collect as much measured data as possible. For this study, the data sources were mainly the HGI-CGS database, unpublished reports, and published articles. Groundwater samples for the analysis of stable isotopes and tritium in groundwater were collected in different periods from 1997 to 2014 (Tables 1 and 2).

The stable isotope compositions were analysed at the Jožef Stefan Institute (Ljubljana, Slovenia) from 1997 to 1999. They were measured on a Varian MAT 250 mass spectrometer [46]. From 2001 to 2005, the measurements were conducted at Joanneum Research (Graz, Austria). An isotope ratio mass spectrometer (IRMS) was used for the measurements. The stable isotope composition of the water from 2005 to 2014 was determined at SILab [Stable Isotope Laboratory at the Physics Department, School of Medicine, University of Rijeka, Rijeka, Croatia]. An HDO Equilibration Unit (ISO Cal, Phoenix, AZ, USA) attached to the dual inlet port of a DeltaPlusXP (Thermo Finnigan, Waltham, MA, SAD) IRMS was used [23].

The results of the stable isotope composition are expressed as the δ-values per mill (‰) relative to the standard: $\delta_{S/R} = (R_{Sample}/R_{Reference}) - 1$ [14,57–59]. R_{Sample} and $R_{Reference}$ mark the isotope ratio ($R = {}^2H/{}^1H$ and $R = {}^{18}O/{}^{16}O$) in the sample. The analytical precision of the determined δ-values was better than ±0.1‰ for $\delta^{18}O$ and ±1‰ for δ^2H.

Tritium activity concentrations were mostly analysed at the Department of Experimental Physics of the Ruđer Bošković Institute (RBI) in Zagreb. The gas proportional counting (GPC) technique was used until 2007 [60,61]. Since 2010, samples have been measured only using the liquid scintillation counting technique following electrolytic enrichment (LSC-EE), while during 2008 and 2009, both techniques were used [47,62–65]. The GPC technique was replaced by the LSC-EE technique for the following reasons: (i) the tritium activity approached natural pre-bomb levels (<5–10 TU), so the measurement of samples without tritium enrichment was not sufficiently precise and (ii) the GPC technique did not satisfy the requirements for a low detection limit and a high throughput of samples. In the RBI reports from 1997 to 1999, the results were expressed in Bq/L and later in TU and Bq/L. The detection limit was 2.5 TU, and the measurement uncertainty was between 2 and 5 TU, depending on the activity concentration [41]. The tritium activity concentrations in some samples collected from 2005 to 2008 were determined at the Isotope Hydrology Section Laboratory at the IAEA [26]. Tritium activity concentrations in the samples collected in 2013 and 2014 were measured at Hydrosys Labor Ltd. in Budapest, Hungary. The analytical method was based on the MSZ 19387:1987 standard. The results are expressed in TU.

For locations featuring a large amount of measured data, the descriptive statistics of stable isotopes and tritium activity concentrations (minimum and maximum values, arithmetic mean, and standard deviation) were used to describe the isotopic data of groundwater variability in the territory of Croatia. In several locations, there was only one measured datum, which was also used in the analysis. Data divided into two categories were used to create the attached diagrams. The term "mean values" is used when the displayed values represent the calculated arithmetic mean of all values measured so far at a particular measuring point. Individual results of all performed measurements were used on diagrams made based on "single values". The first step in the analysis was to compare the δ^2H and δ^{18}O isotopic compositions of groundwater for each site considering the global meteoric water line (GMWL), as well as the local meteoric water lines (LMWLs).

If δ^2H is plotted versus δ^{18}O, the data group form a linear trend line, which can be described by its slope and intercept. The global meteoric water line (GMWL) defines the general relation between δ^2H and δ^{18}O in the precipitation on a global scale and is described by δ^2H = 8 × δ^{18}O + 10 [83–85]. For regional and local areas, the slope and intercept can differ from the GMWL, so for hydrogeological research, the regional/local meteorological water line (RMWL/LMWL) can be more appropriate. A comparison of RMWL/LMWL with GMWL shows local deviations from the world average. A direction coefficient (slope) of LMWL with a value greater than 8 indicates multiple moisture circulation, whereas a direction coefficient of less than 8 indicates greater moisture loss through evaporation [86].

There are different ways to calculate slope and intercept. Ordinary least squares regression (OLSR) is most commonly used for this purpose. Recently, reduced major axis (RMA) regression has been used. These methods do not, however, take into account the amount of precipitation. A new precipitation weighted least square regression (PWLSR) method and precipitation-weighted regressions (PWRMA) method have been applied more recently [87,88]. In this paper, the regional meteoric water lines (RMWL) for the continental and coastal parts of Croatia [89] are used alongside other published LMWLs [20,26,27,41,45–47,90–93] (Table 3).

Table 1. Information on the measurement points in the Pannonian part of Croatia. Legend: measurement point type (OW—observation well, S—spring), lithology (G—gravel, S—sand, L—limestone; D—dolomite).

Sampling Location	Measurement Point Type	Longitude (°)	Latitude (°)	Z (m a.s.l.)	Screen Bottom (m a.s.l.)	Lithology	Geographical Area	Sampling Time	Number of Stable Isotope Samples	Number of ^{3}H Samples	Data Source
ČDP-9/1	OW	16.07	45.76	107.75	31.8	G		2001–2003	7	7	[28,66]
ČDP-9/2	OW	16.07	45.76	107.75	62.8	G		2001–2003	7	7	[28,66]
ČDP-9/3	OW	16.07	45.76	107.75	88.8	G	western	2001–2003	7	7	[28,66]
ČDP-23/1	OW	16.13	45.75	103.79	18.8	G	part	2001–2003	7	7	[28,66]
ČDP-23/2	OW	16.13	45.75	103.79	47.8	G	of the	2001–2003	7	7	[28,66]
ČDP-23/3	OW	16.13	45.75	103.79	86.2	G	Sava basin	2001–2003	7	7	[28,66]
PP-18/20	OW	16.04	45.77	108.25	87.8	G	(Zagreb	2001–2003	7	3	[28,66]
PP-18/40	OW	16.04	45.77	108.25	67.9	G	area)	2001–2003	7	3	[28,66]
PP-18/90	OW	16.04	45.77	108.25	19.3	G		2001–2003	7	3	[28,66]
JP-10	OW	16.04	45.77	107.88	70.1	G		2001–2003	7	4	[28,66]
SPB-J	OW	17.06	45.93	175.00	79.0	G, S	Bilogora Mt.	2013	1	2	[66,67]
GR-1D	OW	16.75	46.21	143.04	99.1	G		2013–2014	7		[66,68]
KP-12A	OW	16.85	46.14	138.50	115.5	G		2013–2014	11	2	[66,68]
KP-12	OW	16.85	46.14	138.00	49.0	G	western	2013–2014	12	2	[66,68]
SO-1	OW	16.95	46.10	126.00	112.0	G, S	part	2013–2014	12	2	[66,68]
O-6	OW	16.95	46.10	126.00	54.0	G, S	of the	2013–2014	12		[66,68]
VIRJE	OW	17.01	46.06	125.00	115.0	G, S	Drava	2013–2014	12		[66,68]
SPB-8	OW	17.05	46.06	117.50	49.5	G, S	basin	2013–2014	11	2	[66,68]
SPB-10	OW	17.03	46.05	126.50	62.5	G, S		2013–2014	7		[66,68]
ÐN-3	OW	17.08	45.98	141.00	92.7	G, S		2013–2014	7		[66,68]
SPB-4	OW	18.47	45.13	83.00	16.3	G, S	Eastern	2012–2013	17		[29,66]
SPB-7	OW	18.45	45.10	85.50	16.2	G, S	Slavonia—Sava	2012–2013	17	1	[29,66]
SPB-9	OW	18.45	45.10	85.00	9.5	G, S	basin	2012–2013	17	1	[29,66]
V-13	OW	18.48	45.07	87.00	55.8	G, S		2012–2013	17	1	[29,66]
SPBPv	OW	16.44	45.27	182.70		L		2006–2007	4	1	[66,69]
B-1	OW	16.42	45.29	178.60		L		2006–2007	4	1	[66,69]
B-2	OW	16.42	45.29	178.60		L		2006–2007	4	1	[66,69]
B-3	OW	16.42	45.29	178.60	145.6	L	Banovina	2006–2007	4	1	[66,69]

Table 1. *Cont.*

Sampling Location	Measurement Point Type	Longitude (°)	Latitude (°)	Z (m a.s.l.)	Screen Bottom (m a.s.l.)	Lithology	Geographical Area	Sampling Time	Number of Stable Isotope Samples	Number of ^{3}H Samples	Data Source
Bojanića vrelo	S	16.43	45.29			L		2006–2007	4	1	[66,69]
Grabovac	S	16.41	45.28			L		2006–2007	4	1	[66,69]
Pašino vrelo	S	16.43	45.29			L		2006–2007	4	1	[66,69]
Veličanka	S	17,64	45,50	512.00		D		2002	1	1	[66,70]
Dubočanka	S	17.68	45.50	490.00		D		2002	1	3	[66,70,71]
Tisovac I	S	17.69	45.50	495.00		D	Papuk Mt.	2002, 2013	1	3	[66,70,71]
Tisovac II	S	17.69	45.50	487.00		D		2002, 2013	1	3	[66,70,71]
Tisovac III	S	17.68	45.49	465.00		D		2002, 2013	1	3	[66,70,71]
Subthermal spring	S	17.66	45.47	287.00		D		2013		2	[66,71]
LO-1	OW	16.06	46.17	322.00	Open hole to 140 m	D		2013		1	[66,72]
LO-4	OW	16.06	46.16	280.00		D	Ivanščica Mt.	2013		1	[66,72]
LO-5	OW	16.06	46.17	305.00	Open hole to 214–244 m	D		2013		1	[66,72]
Škrabutnik	S	16.08	46.19	447.00		D		2013		1	[66,72]

Table 2. Information on the measurement points in the karst part of Croatia. Legend: measurement point type (OW—observation well, S—spring, M—mine).

Sample ID	Measurement Point Type	Longitude (°)	Latitude (°)	Z (m a.s.l.)	Geographical Area	Sampling Time	Number of Stable Isotope Samples	Number of ^{3}H Samples	Data Source
Bubić jama	S	14.16	45.14	20.0		1998	2	2	[66,73]
Bulaž	S	13.89	45.38	18.0		1998, 2001–2003	9	7	[66,73,74]
Fonte Gajo	S	14.07	45.07	5.0		1998	2	2	[66,73]
Gradole	S	13.70	45.34	5.0		1998, 2001–2003	9	7	[66,73,74]
Rakonek	S	14.02	45.09	5.0		1998	2	2	[66,73]
Rudničke vode	M	14.15	45.08	2.0	Istria	1998	2	2	[66,73]
Sv. Ivan	S	13.98	45.40	48.0		1998, 2001–2003	9	7	[66,73,74]
Šišan	S	13.92	44.86	49.0		1998	2	2	[66,73]
Funtana	S	13.61	45.18	2.0		1998	1	2	[66,73]
Karpi	S	13.85	44.93	50.0		1998	1	2	[66,73]
Vela Učka	S	14.19	45.30	926.0		1998	2	2	[66,73]
Rječina	S	14.42	45.42	325.0		1997–1998, 2001–2003, 2012–2013	23	12	[12,66,73,74]
Zvir	S	14.45	45.33	5.0		1997–1998, 2001–2003, 2012–2013	23	13	[12,66,73,74]
Kristal	S	14.30	45.33	0.0		2001–2003	7	7	[66,74]
Novljanska Žrnovnica	S	14.85	45.12	1.5	Hrvatsko Primorje	1997, 1999	6	4	[66,75]
Martinščica	OW *	14.48	45.32	2.0		1999	5	4	[66,75]
Perilo	OW *	14.54	45.31	1.0		1999	4	4	[66,75]
Čabranka	S	14.64	45.60	552.0		1997–1999, 2001–2003	10	10	[66,74,76]
Kupa	S	14.69	45.49	320.0		1997–1999	5	5	[66,76]
Kupica	S	14.85	45.43	250.0		1997–1999	5	5	[66,76]
Kupari	S	14.70	45.50	310.0		1997–1999	5	5	[66,76]
Mala Belica	S	14.80	45.46	290.0		1997–1999	5	5	[66,76]

Table 2. *Cont.*

Sample ID	Measurement Point Type	Longitude (°)	Latitude (°)	Z (m a.s.l.)	Geographical Area	Sampling Time	Number of Stable Isotope Samples	Number of ^{3}H Samples	Data Source
Velika Belica	S	14.76	45.48	290.0	Gorski Kotar	1997–1998	4	4	[66,76]
Zeleni vir	S	14.90	45.42	350.0		1997–1999	5	5	[66,76]
Gerovčica spring	S	14.68	45.53	320.0		1997–1999	5	5	[66,76]
Gomirje	S	15.13	45.33	377.0		2002	1		[66]
Mala Lešnica	S	14.85	45.44	238.0		2002	1		[66]
Vitunjčica	S	15.14	45.29	349.0		2002	1		[66]
Kamačnik	S	15.06	45.35	415.0		2002	1		[66]
Majerovo vrelo	S	15.36	44.81	462.0		1996, 2005–2008	28	21	[26,66,77]
Tonkovića vrelo	S	15.37	44.79	457.0		1996, 2005–2008	27	18	[26,66,77]
Pećina	S	15.33	44.80	456.0		1996, 2005–2008	29	21	[26,66,77]
Klanjac	S	15.37	44.79	458.0	Lika (Adriatic Sea basin)	2005, 2008	14		[26,66]
Grab	S	15.32	44.85	454.0		2008	10		[26,66]
Jaz	S	15.32	44.81	456.0		2005, 2008	10		[26,66]
Knjapovac	S	15.36	44.79	459.0		2005–2006, 2008	13		[26,66]
Marusino vrelo	S	15.32	44.81	456.0		2005–2006, 2008	14		[26,66]
Korenička Rijeka spring	S	15.65	44.78	695.0		2006	1	1	[66,78]
Kameniti vrelac	S	15.66	44.77	690.0		2006	1	1	[66,78]
Koreničko vrelo	S	15.67	44.77	700.0	Lika (Danube River basin)	2006	1	1	[66,78]
Makarevo vrelo	S	15.66	44.78	695.0		2006	1	1	[66,78]
Mlinac	S	15.66	44.78	700.0		2006	1	1	[66,78]
Stipinovac	S	15.46	44.97	696.0		2006	1	1	[66,78]

Table 2. *Cont.*

Sample ID	Measurement Point Type	Longitude (°)	Latitude (°)	Z (m a.s.l.)	Geographical Area	Sampling Time	Number of Stable Isotope Samples	Number of ³H Samples	Data Source
Malo vrelo Ličke Jesenice	S	15.44	44.97	481.0		2007–2008	3	3	[66,79]
Veliko vrelo Ličke Jesenice	S	15.46	44.97	479.0		2007–2008	3	3	[66,79]
Vrilo Velebita	S	15.54	44.26	0.0		1999	1		[66,80]
Vrulja Modrič	S	15.54	44.26	0.0		1999	1		[66,80]
Jadro	S	16.52	43.54	40.0		2001	4	4	[66,81]
Žrnovica	S	16.57	43.52	90.0		2001	4	4	[66,81]
B-22A	OW	16.71	43.58	335.0	Dalmatia	2001	4	4	[66,81]
Krka	S	16.24	44.04	225.0		2001	1	1	[66,82]
Lopuško vrelo	S	16.22	44.02	230.0		2001	1	1	[66,82]
Kosovčica spring	S	16.25	43.94	250.0		2001	1	1	[66,82]
Modro oko	S	17.51	43.06	1.5		2005	2		[66]
Opačac	S	17.18	43.45	266.2		2005	2		[66]

* group of observation wells.

Table 3. Local meteoric water lines (LMWLs) used in this paper.

Name of RMWL/LMWL in This Paper	Station (Altitude m a.s.l.)	Observation Period	δ^2H/δ^{18}O Correlation Equation	R or rmSSE *	N	Data Source
RMWL_continent	Bilogora (202), Zagreb (158), Sirač (161), Plitvička jezera (580), Sikirevci (82), Veliki Grđevac (148), Medicinska škola Varaždin (173), Sokolovac (207), Jastrebarsko (152), Duga Resa (130), Virovitica (122), Slavonski Brod (92), Bedekovčina (160), Sošice (560), Pogana jama (980), Grdanjci (185) i Karlovac (126)	2008–2013	δ^2H = (7.4 ± 0.005) × δ^{18}O + (4.1 ± 0.5)	0.99	524	[89]
RMWL_coast	Labin (263), Vežica (219), Zadar (2), Opatija (0), Kastav (333), Rijeka_Medicinski fakultet (12), Kukuljanovo (281), Ponikve (20), Ičići (1), Vrh, Pehlin (278), Jadranovo (0), Kraj (72), Mali Lošinj (1.9), Zadar GNIP (5), Komiža-Vis GNIP (6), Dubrovnik GNIP (52), Split (42) i Murter (16.6)	2008–2013	δ^2H = (7.0 ± 0.08) × δ^{18}O + (4.4 ± 0.5)	0.96	655	[89]
LMWL_Portorož	Portorož—Airport (2)	2001–2003	δ^2H = (7.7 ± 0.4) × δ^{18}O + (7.3 ± 2.2)	0.96	35	[45]
LMWL_Portorož_OLSR	Portorož—Airport (2)	2000–2006	δ^2H = (7.82 ± 0.23) × δ^{18}O + (7.84 ± 1.57)	0.97	74	[91]
LMWL_Portorož_RMA	Portorož—Airport (2)	2000–2006	δ^2H = (8.05 ± 0.22) × δ^{18}O + (9.35 ± 1.55)	0.97	74	[91]
LMWL_Portorož_PWLSR	Portorož—Airport (2)	2007–2010	δ^2H = (7.8 ± 0.27) × δ^{18}O + (8.52 ± 1.85)	0.96	71	[92]
LMWL_Hrvatsko Primorje	Kukuljanovo (281), Pehlin (278), Škalnica (526), Gumance (688), Ilirska Bistrica (1043), Snežnik (1300)	2010–2012	δ^2H = (8.2 ± 0.1) × δ^{18}O + (14.0 ± 1)	0.98	88	[25]
LMWL_Zavižan	Zavižan—Velebit Mt. (1594)	2001–2003	δ^2H = (7.6 ± 0.4) × δ^{18}O + (10.5 ± 4.0)	0.95	35	[45]
LMWL_Gacka	Ličko Lešće (457), Gospić (656)	2005–2006	δ^2H = 6.8 × δ^{18}O + 1.5			[23]
LMWL_Ljubljana	Ljubljana (299)	2001–2003	δ^2H = (8.0 ± 0.2) × δ^{18}O + (9.2 ± 1.8)	0.99	36	[45]
LMWL_Ljubljana (Bežigrad)_OLSR	Ljubljana (299)	1981–2006	δ^2H = (7.95 ± 0.08) × δ^{18}O + (8.9 ± 0.71)	0.99	290	[46]

Table 3. *Cont.*

Name of RMWL/LMWL in This Paper	Station (Altitude m a.s.l.)	Observation Period	$\delta^2H/\delta^{18}O$ Correlation Equation	R or rmSSE *	N	Data Source
LMWL_Ljubljana (Bežigrad)_RMA	Ljubljana (299)	1981–2006	$\delta^2H = (8.06 \pm 0.08) \times \delta^{18}O + (9.84 \pm 0.71)$	0.99	290	[46]
LMWL_Ljubljana (Reaktor)_RMA	Ljubljana	2007–2010	$\delta^2H = (8.19 \pm 0.22) \times \delta^{18}O + (11.52 \pm 1.97)$	0.98	46	[47]
LMWL_Ljubljana (Bežigrad)_PWLSR	Ljubljana (299)	2007–2010	$\delta^2H = (7.94 \pm 0.21) \times \delta^{18}O + (9.76 \pm 1.91)$	0.99	46	[47]
LMWL_Zagreb	Zagreb (157)	2001–2003	$\delta^2H = (7.3 \pm 0.2) \times \delta^{18}O + (2.8 \pm 1.8)$	0.99	34	[45]
LMWL_Zagreb_OLSR *	Zagreb (157)	1980–2018	$\delta^2H = (7.65 \pm 0.06) \times \delta^{18}O + (4.79 \pm 0.55)$	1.0047	389	[41]
LMWL_Zagreb_RMA *	Zagreb (157)	1980–2018	$\delta^2H = (7.74 \pm 0.06) \times \delta^{18}O + (5.57 \pm 0.55)$	1.0019	389	[41]
LMWL_Zagreb_PWLSR *	Zagreb (157)	1980–2018	$\delta^2H = (7.64 \pm 0.06) \times \delta^{18}O + (5.24 \pm 0.54)$	1.00	389	[41]
LMWL_Bilogora	Bilogora (202)	2010–2013	$\delta^2H = 7.75 \times \delta^{18}O + 8.11$	0.99	19	[93]
LMWL_Sikirevci	Sikirevci (85)	2012–2014	$\delta^2H = 7.69 \times \delta^{18}O + 6.29$	0.99	21	[29]

* rmSSE: average of the root mean square sum of squared errors [85].

Most LMWLs are defined by the least squares regression method; the exceptions are the LMWLs for Ljubljana and Portorož [47,91,92] and Zagreb [41]. Previous studies of different approaches to calculate the LMWL have led to the conclusion that all regression lines have similar values for both the slope and the intercept [41,47,91,92]. This indicates that the LMWLs defined by the OLSR method can be used for comparisons with the measured isotopic composition of groundwater in Croatian regions. The absence of a significant difference between the PWLSR slope and both the OLSR and RMA slopes indicates a relatively homogeneous distribution of monthly rainfall, as well as little small monthly rainfall with a minor excess of deuterium [41,88].

Deuterium excess (*d*-excess) is defined as $d\,(‰) = \delta^2H - 8 \times \delta^{18}O$ [83]. This is an isotope parameter that indicates the deviation of local samples from the GMWL and is an indicator of climate sensitivity at the source of humidity, as well as along the trajectory of air masses into the atmosphere [94]. In other words, *d*-excess reflects the prevailing conditions during the evolution and interaction or mixing of air masses en route to the precipitation site [86].

Tritium activity concentrations in the groundwater samples were used to estimate an approximate mean groundwater age (MRT). The qualitative interpretations in this study are made as follows [9]: <0.8 TU indicates sub-modern water (recharged prior to 1950s), 0.8 to ≈ 4 TU indicates a mix of sub-modern and modern water, 5 to 15 TU indicates modern water, 15–30 TU indicates some "bomb" tritium, and >30 TU indicates that a recharge occurred in the 1960s to 1970s.

3. Results and Discussion

The statistical isotopic data of groundwater samples from the observation wells and springs are presented in Tables S1a, S2a, S3a and S4a. On the sampling sites where the composition of stable isotopes was measured only once, groundwater isotopic data are shown in Tables S1b, S2b, S3b and S4b. The tables can be found in Supplementary Files.

3.1. Stable Isotope Composition of Groundwater in the Karst Area of Croatia

The mean $\delta^{18}O$ values in the drinking karst groundwater range from −10.9‰ to −5.5‰. Similarly, the lowest mean δ^2H was −71.9‰, and the highest δ^2H was −34.0‰ (Table S1a, Figure 2a). The most positive δ-values were measured in the spring below the sea surface (submarine spring, vrulja Modrič), reflecting the influence of sea water (Figure 2b). A slightly more negative value of stable isotopes was measured in the groundwater sample of the coastal spring Kristal in the summer of 2001 (Figure 2b). Since the sampling was done in summer, during the dry season, the influence of the sea was very pronounced, as observed by the isotopic composition of the groundwater. The $\delta^{18}O/\delta^2H$ values for all karst springs are distributed along the regional meteoric water line of the coastal part of Croatia (RMWL_cost in Table 3) and indicate a meteoric origin. This also applies to springs belonging to the Danube River basin that are approximately 50 km away from the sea. Consequently, most precipitation that exerts an influence on groundwater recharge in the karst area comes from the precipitation supplied by the south wind in the direction of the Adriatic sea (more often in the cold than in the warm part of the year).

A comparison of RMWL with GMWL shows local deviations from the world average. The slope of RMWL_coast has a value less than 8 (Table 3), indicating greater moisture losses through evaporation. Evaporation moisture losses occur due to low rainfall, during very hot summers, or for both reasons simultaneously [95], which is characteristic of this part of Croatia.

The $\delta^{18}O/\delta^2H$ values form the regional karst groundwater line (RGWL_karst), which is described by following equations—$\delta^2H = 6.8 \times \delta^{18}O + 2.5$ (n = 38, $R^2 = 0.98$), for the mean values of the stable isotopes (Figure 2a), and $\delta^2H = 7.0 \times \delta^{18}O + 4.2$ (n = 340, $R^2 = 0.96$), for the single values of the stable isotopes (Figure 2b). Both lines have very similar slopes and intercepts to RMWL_coast.

Depending on their affiliation with their geographic areas, individual groups of springs are clearly located along the RMWL_coast. However, some spring groups fit better with LMWLs. The single values of stable isotopes of karst groundwater from Figure 2b and LMWLs are shown in Figure 3.

Figure 3a shows the isotopic composition of the spring water in Istria and the LMWLs defined by Vreča et al. [45,91,92] (Table 3). The $\delta^{18}O/\delta^2H$ data of the spring water fit very well with the LMWLs determined by the isotopic precipitation composition in Portorož, Slovenia. This is unsurprising, since Portorož is less than 50 km away from central Istria and climatically belongs to the same area. The figure shows several LMWLs that are determined using different methods. All regression lines have similar values (within uncertainties) for both the slope and the intercept values [92]. The slopes of LMWLs obtained using data for the period 2007–2010 and the PWLSR method are close to those of the LMWLs using the OLSR and RMA method for the periods of 2001–2003 and 2000–2006.

The majority of the springs that belong to the Gorski Kotar (Kupica, Mala Belica, Gerovčica, Vela Učka, Zvir, Velika Belica, Kupa, and Kupari) in the Danube River basin and Hrvatsko Primorje (Vela Učka, Rječina, Zvir Perilo, and Martinščica) in the Adriatic Sea basin (Table 2) are situated in the zone between approximately $\delta^{18}O = -8.6‰$ and $\delta^{18}O = -7.9‰$, as well as $\delta^2H = -55.9‰$ and $\delta^2H = -48.9‰$ (Figure 3b). Although some of these springs are located along the coast (Zvir, Perilo, and Martinscica) and some at several hundred meters above sea level (Rječina, Vela Učka, Kupica, Mala Belica, Gerovčica, Velika Belica, Kupa, and Kupari), their $\delta^{18}O$ values have small variations. Their isotopic compositions are mainly influenced by the climatic conditions in the recharge area of the aquifer, which are approximately the same (found in close proximity to one another) for the considered springs.

The mean residence time (MRT) of groundwater was calculated for the stable isotope $\delta^{18}O$ using the lumped parameter approach and applying the exponential model, the combined exponential-piston model, and the dispersion model to the isotopic input (rainfall) and output (spring) datasets during 2011–2013 [12]. The MRTs of 3.24 and 3.6 months for the Rječina spring and 7.2 months for the Zvir spring suggest the presence of recent groundwater recharge from precipitation, as well as fast groundwater flow. The cumulative age distributions show that the proportion of water younger than 1xMRT at both springs was more than 50% and that the proportion of water younger than 2xMRT was almost 90%.

The high mean δ-values $\delta^{18}O = -5.9‰$ to $-5.2‰$ and $\delta^2H = -37.6‰$ to $-32.5‰$ in Hrvatsko Primorje refer to the coastal spring Kristal in Opatija (Figures 1 and 3b). The water is brackish, and during low water hydrological conditions, the proportion of seawater is considerable, which is reflected in the composition of the stable isotopes in the water.

Figure 3c shows the isotopic composition of the spring water in Gorski Kotar and Lika, as well as the LMWLs defined by Vreča et al. [45] and Mandić et al. [23] (Table 3). The figure clearly shows that the LMWL defined by Mandic et al. [23] fits well with groundwater samples at all Lika springs and at the springs located in the eastern part of Gorski Kotar. At the same time, LMWL_Zavižan [45] fits better with the springs in the westernmost part of Gorski Kotar, whose recharge area is situated at a high altitude, as well as the Zavižan.

The lowest δ-values were recorded on the springs that belong to the Lika area in the Danube River basin (Table 2, Figure 3c). Although these springs are located in the Danube River basin, it is evident that aquifer recharge is carried out by precipitation, which is influenced by the Mediterranean air mass. In this area (but east of Malo vrelo Ličke Jesenice and Veliko vrelo Ličke Jesenice), the composition of the stable isotopes was once measured at six springs (Korenička Rijeka spring, Kameniti vrelac, Koreničko vrelo, Makarevo vrelo, and Mlinac i Stipinovac) in the Una river basin (Table 2, Figure 1). The measured $\delta^{18}O$ values were in the range of $-10.9‰$ to $-10.2‰$ and $-71.9‰$ to $-67.1‰$ for δ^2H (Table S1b), which is very similar to the mean isotopic composition of Malo vrelo Ličke Jesenice and Veliko vrelo Ličke Jesenice.

At the same time, at springs that belong to the Lika area in the Adriatic Sea basin (Majerovo vrelo, Klanjac, Tonkovića vrelo, Grab, Marusino vrelo, Jaz, and Knjapovac—in Table 2), the mean $\delta^{18}O$ values were in the range of $-10.2‰$ to $-8.0‰$, as well as $-71.0‰$ to $-50.0‰$ for δ^2H (Table S1a). All springs, including those in the Danube River basin and those in the Adriatic Sea basin, are situated at similar altitudes (Table 2), so the differences between the stable isotope compositions of the springs belonging to the Danube River basin and the springs belonging to the Adriatic Sea basin could be caused by the "continental effect", also referred to as the distance-from-coast effect [39]. However, in this case, the distance from the sea is likely not the reason because the differences are very small, (approximately 20 kilometres) (Figure 4). Although the springs are at similar altitudes, the altitudes of their recharge areas are different. Higher altitudes in the catchment area of the springs in the Danube River basin likely have a much greater impact than the distance from the coast. As the altitude of the terrain increases, the $\delta^{18}O$ and δ^2H in precipitation becomes increasingly more depleted. This effect correlates to air temperature, which drops due to an increase in altitude [39]. The $\delta^{18}O$ effect generally varies between −0.1‰ and −0.6‰/100 m of altitude and often decreases with an increase in altitude [96].

The mean $\delta^{18}O$ values of spring water in Dalmatia vary between −6.7‰ and −8.7‰ and −40.0‰ and −58.0‰ for δ^2H (Figure 3d), which is, according to Kapelj et al. [21], within the range of the stable isotope composition of groundwater in middle Dalmatia. The composition of stable isotopes in groundwater, especially in the spectrum of depleted values, shows similarity to LMWL_Zavižan [45] and thus reflects the influence of the recharge area at high altitudes (e.g., Zavižan—Velebit Mt. 1594 m a.s.l., Dinara Mt. 1830 m a.s.l.). However, the stable isotopes data of measured groundwater define a local groundwater line (LGWL_Dalmatia) that better matches the RMWL_coast (Figure 3d). It is described using equation $\delta^2H = 6.6 \times \delta^{18}O + 1.7$ (n = 21, R^2 = 0.97). The difference between the slope values of the lines for RGWL_coast and LGWL_Dalmatia is generally about 0.42‰ and between the axis intercept values is about 2.74‰. These differences between LMWL_Zavižan and LGWL_Dalmatia are 1.02‰ for slope values and 8.84‰ for intercept values. Kapelj et al. [21] analysed the mean altitude of the spring catchment areas and found the $\delta^{18}O$ gradient to be between −0.2‰ and −0.4‰/100 m, which corresponds to the value of −0.3‰/100 m determined by Vreča et al. [46] using stable isotope data for precipitation at several stations in Croatia and Slovenia, as well as that of Mandić et al. [23], who used stable isotope data for the groundwater at Lika springs Majerovo vrelo, Tonkovic vrelo, and Pećina. Using data on the stable isotopes in precipitation at several stations in the Učka area (Rijeka hinterland), Hunjak [89] determined a $\delta^{18}O$ gradient of −0.19‰/100 m. However, these relationships cannot be applied to all springs analysed in this paper because they have different catchment areas that differ in their temperature changes and wind directions, which both significantly affect the composition of stable isotopes in precipitation and thus in groundwater.

The *d*-excess values of karst groundwater were determined to be between 6‰ and 17‰ (Table S1, Figure 5). *d*-values lower than 6‰ were found in three individual samples (coastal Kristal spring, vrulja Modrič, and Šišan spring in Istria), which are under the influence of sea water. In many cases, *d*-excess is found to increase with altitude [39,45]. Atlantic air masses typically have *d*-excess values around 10‰, while Mediterranean air masses are characterized by higher *d*-excess values of approximately 12‰ [14]. The catchment areas of the springs in Lika, Gorski Kotar, and Hrvatsko Primorje are at relatively high altitudes, and due to the influence of high altitudes and Mediterranean air masses, their *d*-excess values are high and reach 17‰ (Figure 5b). The lowest *d*-excess values are determined in springs in Istria whose catchment areas are at relatively low altitudes and are partly affected by Atlantic air masses.

3.2. Stable Isotope Characteristics of Groundwater in the Pannonian Area of Croatia

$\delta^{18}O$ values of groundwater ranged from −11.0‰ to −8.2‰ and from −76.0‰ to −57.0‰ for δ^2H (Table S2, Figure 6). The lowest $\delta^{18}O$ and δ^2H values were measured in the spring water on Papuk Mt (Figure 6b). As an increase in altitude leads to depleted isotope values, it is expected that among all the analysed groundwater samples, these springs will have the lowest isotopic values as their recharge areas are situated at the highest altitudes, reaching 825 m a.s.l. The highest values of stable isotopes were measured in groundwater in the Zagreb area where they range from −9.5‰ to −8.2‰ for $\delta^{18}O$ and from −65.0‰ to −57.0‰ for δ^2H (Figure 6b). The altitude of the groundwater recharge in this area reaches up to 200 m a.s.l. As shown in Figure 6b, the stable isotope values of groundwater samples from Eastern Slavonia—the Sava basin, Banovina, and the Drava River valley—are located between the stable isotope values of groundwater on Papuk Mt. and the Zagreb area. The groundwater recharge areas in the Drava river valley reach approximately 250 m a.s.l. and 290 m a.s.l. in the Banovina area. The catchment area of groundwater captured by observation wells in Eastern Slavonia—the Sava basin—is mostly north of the wells [29], possibly also on the slopes of the Slavonian highlands up to altitudes of about 250 m a.s.l.

The groundwater isotopic values in the Pannonian part of Croatia lie slightly above the RMWL_continent and indicate a meteoric origin (Figure 6). Considering that only a small fraction of precipitation actually reaches groundwater in these types of aquifers, the meteoric signal in groundwater is muted. The regional groundwater line (RGWL_Pannon) is developed from groundwater $\delta^{18}O/\delta^2H$ values. It is described by the following equations—$\delta^2H = 7.1 \times \delta^{18}O + 2.3$ (n = 30, R^2 = 0.97), for the mean values of the stable isotopes (Figure 6a), and $\delta^2H = 7.4 \times \delta^{18}O + 5.5$ (n = 255, R^2 = 0.93), for the single values of the stable isotopes (Figure 6b).

As in the case of karst groundwater, the groundwater in certain areas in the Pannonian part of Croatia is compared with LMWLs (Figure 7a–d). The single values of stable isotopes of groundwater in the Pannonian part of Croatia from Figure 6b and LWWLs are shown in Figure 7. The isotopic composition of the studied groundwater in the Zagreb area fits better with LMWL_Ljubljana [45–47] than with LMWL_Zagreb [41,45]. This was previously identified by Marković et al. [24], Horvatinčić et al., [27], and Parlov et al. [30]. The different methodologies used to calculate the LMWLs for Ljubljana [45–47] showed slightly better matching of the groundwater isotopic composition with the LMWL determined using the PWLSR method (Figure 7b). Considering that the precipitation weighted least square regression (PWLSR) method takes into account the precipitation amount and that the analysed groundwater is relatively close to the surface of the terrain and to the Sava River (which has an impact on groundwater recharge), the result is unsurprising. The values of the stable isotopes are scattered and show seasonal influence, which points to a relatively short mean residence time of the groundwater in this area. The aquifer is directly exposed to precipitation because there are no covering deposits, as well as to the influence of the Sava River because the observation wells are located in its immediate vicinity. Figure 7c shows the composition of the stable isotopes in the groundwater in the Drava River valley. The LMWL was developed on the basis of the measured values of the stable isotopes on Bilogora Mt. [93], south of the observed wells. The isotopic composition of the groundwater lies slightly above the LMWL_Bilogora. Stable isotopes were analysed over a wide area and at different depths of the aquifer (Table 1), and their values are grouped according to these elements (Figure 7c). The depleted isotope composition is characteristic of groundwater from the deeper parts of the aquifer system and the enriched isotopes are characteristic of the shallower parts of aquifers.

Figure 7d shows the composition of the stable isotopes of groundwater in the aquifer system in Eastern Slavonia—the Sava basin—as well as the LMWL that was made on the basis of the stable isotope values of the precipitation, which were measured in the immediate vicinity of the analysed wells [29]. The stable isotope composition data in groundwater are slightly above the LMWL and are grouped into almost a single point. Since groundwater is accumulated in the aquifer system in which the mean residence time is several decades (as evidenced by ^3H activity concentrations), the water is homogenized so that the differences in the stable isotope composition of the precipitation—which recharge the groundwater—completely disappear.

The mean *d*-excess values of groundwater were determined to be between 8.7‰ and 12.1‰ (Table S2, Figure 8a), while the single values range from 7.8‰ to 14.4‰ (Figure 8b). Higher values of *d*-excess are attributed to the influence of Mediterranean precipitating air masses [41,45]. Bottyán et al. [97] estimated that the moisture sources of precipitation in Hungary come from the Mediterranean region (57.0%), local moisture (14.8%), the Atlantic region (14.2%), Northern Europe (7.4%), and Eastern Europe (6.6%). Most of the *d*-excess values in the groundwater samples range from 10‰ to 12‰. This is especially pronounced in the groundwater samples in Eastern Slavonia—the Sava basin—, where the mean residence time of the groundwater is relatively long [29] and, accordingly, the range of *d*-excess values is relatively narrow—i.e., the differences in climatic conditions (the source of humidity and contribution of air masses) during aquifer recharge are muffled in the groundwater samples. Opposite to this, *d*-excess is found in a wide range of values in the analysed groundwater samples in the Zagreb and Banovina area. Since higher values of *d*-excess are characteristic for winter months and lower values for summer months [14,41,45] (as evident in the groundwater samples), it can be concluded that the mean residence time of groundwater is relatively short.

3.3. Tritium Activity in Groundwater

Spatial distribution of an approximate groundwater age estimated by tritium is shown in Figure 9. The mean ^{3}H activity concentrations in the groundwater mostly vary between 4 TU and 10 TU (Tables S3 and S4). In karst aquifers, the ^{3}H activity of the groundwater in the rainy seasons and the mean tritium activity concentrations indicate modern water [9]. For the karst springs Pećina, Majerovo vrelo, and Tonkovića vrelo (areas of the Lika in the Adriatic Sea basin), the groundwater age was estimated using the ^{3}H activity concentration in groundwater and the lumped parameter model [26]. A similar MRT was defined for all three springs and was estimated to be approximately 12 years.

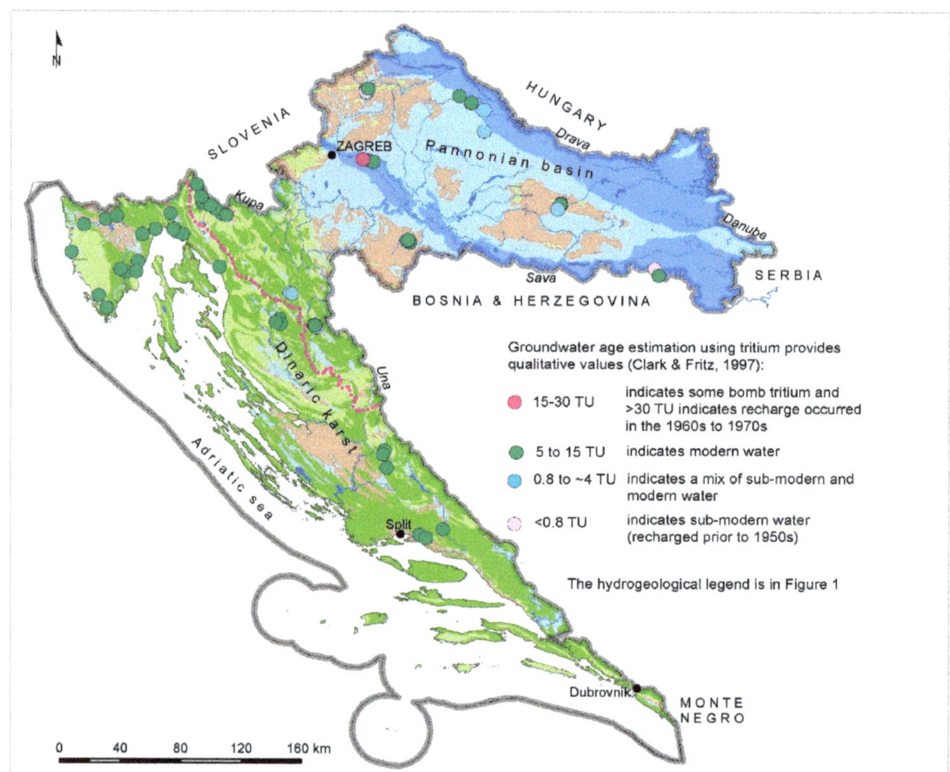

Figure 9. Spatial distribution of an approximate groundwater age using tritium activity.

Modern water is also indicated in the groundwater from alluvial aquifers close to the Sava River (Table S4). The highest ^{3}H activity concentrations were measured in the groundwater in the Zagreb area, as well as in the Sava River, which flows through Zagreb. This result is consistent with the well-known fact that the aquifer is recharged by the Sava River in the Zagreb area [29,52,98,99]. The observation wells are located close to the Sava River. The high ^{3}H activity concentrations in the groundwater samples are explained by the release of tritiated water from Krško Nuclear Power Plant, 30 km upstream from Zagreb [27,100]. This indicates the infiltration of surface water from the Sava

River into the Quaternary alluvial aquifer. Modern water is indicated in the groundwater accumulated in the porous aquifer and observed close to the Sava River in Eastern Slavonia—the Sava basin—during high water periods (Table 2, Figure 9). In the shallow part of the aquifer system in the western part of the Drava basin (Table 1), modern groundwater is also indicated (Table S4).

The values between 0.8 and 4 TU indicate a mix of sub-modern and modern water. These values were measured in the karst aquifers during summer under base flow conditions (Table S3). A mix of sub-modern and modern water is also indicated in the deep porous aquifer of Bilogora Mt. (label 2 in Figure 1), as well as in the deeper part of the alluvial aquifer system in the western part of the Drava basin (Table S4). ^{3}H activity concentrations less than 0.8 TU were measured in the deep alluvial aquifer in Eastern Slavonia as well as in the deep carbonate aquifer on Ivanščica Mt and in the thermal spring on Papuk Mt (Table S4). In Eastern Slavonia—the Sava basin—, but approximately one km away from the Sava river, ^{3}H activity concentrations were measured only once under conditions of high water and accurately reflects the dynamics of groundwater flow in a semi-confined aquifer. Groundwater is recharged north of the sampled observation well and through relatively thick covering deposits, so the mean residence time of groundwater is relatively long [29]. The aquifer is recharged from the Sava River only in its immediate vicinity and only during high water [29], as evidenced by the measurement of tritium activity.

4. Conclusions

This paper presents an analysis of stable isotope data and tritium activity concentrations in Croatian groundwater at approximately 100 sites. Data were collected between 1997 and 2014. The stable isotope composition was compared with different regional and local meteoric water lines. The primary findings of the study are as follows:

- The composition of groundwater stable isotopes in the Pannonian area of Croatia fits well with the regional meteoric water line for the continental portion of Croatia (RMWL_continent), based on which we conclude that RMWL_continent can be accepted for all future research on the composition of stable isotopes in northern Croatia.
- Although the regional meteoric water line for the coastal part of Croatia (RMWL_coast) was developed according to stable isotopic compositions of precipitation, especially at measuring sites located on the Adriatic coast, it is evident that the line fits well with the composition of the stable isotopes of all analysed karst groundwater. Thus, we conclude that the recharge of the karst aquifers is mainly driven by precipitation introduced by winds blowing from the south of the region as shown by the *d*-excess, which has a strong Mediterranean influence.
- The composition of the groundwater stable isotopes in the karst part of Croatia agrees well with RMWL_coast, based on which we conclude that RMWL_coast can be used as an accurate reference for all future research on the composition of stable isotopes in the karstic part of Croatia.
- The composition of stable isotopes in the Pannonian part of Croatia forms a unique regional line of Pannonian groundwater.
- The composition of stable isotopes in the karst part of Croatia forms a unique regional line of karst groundwater.
- This article offers a unique summary of data measured over a long period of time and provides a large picture of the stable isotope data found in Croatia. Especially valuable are the insights into the correspondence between the stable isotope data of precipitation and groundwater through LMWLs and local groundwater lines over a large span of time. The fact that this correspondence is best in the karst area is perhaps not a great surprise (since karst areas are known to have the shortest mean residence time), but that the Mediterranean influence is spreading far beyond mountains of a high altitude and corresponds strongly with isotopic groundwater data are findings that will help in further research.

- According to its isotopic composition, almost all the investigated groundwater used in Croatia for various purposes, primarily as drinking water, is modern water, making such water relatively vulnerable to the potential sources of pollution found in the environment. The groundwater that is accumulated in semi-confined aquifers in eastern Croatia is in a better position, as well as the groundwater in the deep carbonate aquifers in northern Croatia, whose age can reach several decades.

- The number of geographical and temporal data points is very large (approximately 100 groundwater measurement locations and 50 precipitation measurement locations, as well as approximately 1000 measurements of the stable isotopes and tritium activity concentrations in groundwater and more than 1000 measurements of the stable isotopes in precipitation), which is significant for all conclusions made.

The present analysis of data on the isotopic composition of groundwater may be useful for future comparisons with other data in different or identical locations to obtain better knowledge on the isotopic composition of groundwater in Croatia and in nearby countries.

Supplementary Materials: The following are available online at http://www.mdpi.com/2073-4441/12/7/1983/s1 and include isotope data in Tables S1–S4.

Author Contributions: Conceptualization, Ž.B.; analysis and interpretation, Ž.B. and M.K.; writing—original draft preparation, Ž.B., M.K. and T.H.; writing—review and editing, O.L. All authors have read and agreed to the published version of the manuscript.

Funding: This research received no external funding.

Acknowledgments: The authors wish to express their gratitude to the former and current HGI-CGS employees who collected the water samples, as well as the authors of unpublished technical reports used in drafting this paper. Special thanks to Zvjezdana Roller-Lutz, under whose leadership SILab, the first laboratory for measuring stable isotopes in Croatia, was established. A lot of measurements used in this work were made under her guidance and supervision, so the authors thank her for mentorship, useful discussions, and lifelong friendship. The authors wish to thank anonymous reviewers and guest editors for their very useful comments which have helped to improve the manuscript.

Conflicts of Interest: The authors declare no conflicts of interest.

References

1. International Atomic Energy Agency (IAEA). Global Network of Isotopes in Precipitation. The GNIP Database. Available online: https://nucleus.iaea.org/wiser (accessed on 10 July 2020).
2. Maloszewski, P.; Zuber, A. Determining the turnover time of groundwater systems with the aid of environmental tracers. 1. Models and their applicability. *J. Hydrol.* **1982**, *57*, 207–231. [CrossRef]
3. Maloszewski, P.; Zuber, A. Manual on lumped parameter models used for the interpretation of environmental tracer data in groundwater. In *Use of Isotopes for Analysis of Flow and Transport Dynamics in Groundwater Systems*; Yutsever, Y., Ed.; IAEA: Vienna, Austria, 2002; pp. 1–50.
4. Maloszewski, P.; Rauert, W.; Stichler, W.; Herrmann, A. Application of flow models in an alpine catchment area using tritium and deuterium data. *J. Hydrol.* **1983**, *66*, 319–330. [CrossRef]
5. Maloszewski, P.; Rauert, W.; Trimborn, P.; Herrmann, A.; Rau, R. Isotope hydrological study of mean transit times in an alpine basin (Wimbachtal, Germany). *J. Hydrol.* **1992**, *140*, 343–360. [CrossRef]
6. Maloszewski, P.; Stichler, W.; Zuber, A.; Rank, D. Identifying the flow systems in a karst-fissured-porous aquifer, the Schneealpe, Austria, by modelling of environmental ^{18}O and ^{2}H isotopes. *J. Hydrol.* **2002**, *256*, 48–59. [CrossRef]
7. McGuire, K.J.; DeWalle, D.R.; Gburek, W.J. Evaluation of mean residence time in subsurface waters using oxygen-18 fluctuations during drought conditions in the mid-Appalachians. *J. Hydrol.* **2002**, *261*, 132–149. [CrossRef]
8. Rozanski, K.; Dulinski, M. A reconnaissance isotope study of waters in the karst of the Western Tatra mountains. *Catena* **1988**, *15*, 289–301. [CrossRef]
9. Clark, I.D.; Fritz, P. *Environmental Isotopes in Hydrogeology*; Lewis Publishers: London, UK, 1997; p. 328.

10. Maloszewski, P.; Moser, H.; Stichler, W.; Bertleff, B.; Hedin, K. Modelling of groundwater pollution by riverbank filtration using oxigen-18 data. *Ground Water Monit. Manag.* **1990**, *173*, 153–161.

11. Stichler, W.; Maloszewski, P.; Moser, H. Modelling of river water infiltration using oxigen-18 data. *J. Hydrol.* **1986**, *83*, 355–365. [CrossRef]

12. Brkić, Ž.; Kuhta, M.; Hunjak, T. Groundwater flow mechanism in the well-developed karst aquifer system in the western Croatia: Insights from spring discharge and water isotopes. *Catena* **2018**, *161*, 14–26. [CrossRef]

13. Lauber, U.; Goldscheider, N. Use of artificial and natural tracers to assess groundwater transit-time distribution and flow systems in a high-alpine karts system (Wetterstein Mountains, Germany). *Hydrogeol. J.* **2014**, *22*, 1807–1824. [CrossRef]

14. Rozanski, K.; Araguás-Araguás, L.; Gonfiantini, R. Isotopic pattern in modern global precipitation. In *Climate Change in Continental Isotopic Record*; Swart, P.K., Lohwan, K.L., Mckenzie, J., Savin, S., Eds.; American Geophysical Union: Washington, DC, USA, 1993; pp. 1–36.

15. Rozanski, K.; Araguás-Araguás, L. Spatial and temporal variability of stable isotope composition of precipitation over the South American continent. *Bull. Inst. Fr. Études Andin.* **1995**, *24*, 379–390.

16. Aggarwal, P.K.; Alduchov, O.A.; Froehlich, K.O.; Araguás-Araguás, L.J.; Sturchio, N.C.; Kurita, N. Stable isotopes in global precipitation: A unified interpretation based on atmospheric moisture residence time. *Geophys. Res. Lett.* **2012**, *39*. [CrossRef]

17. Sánchez-Murillo, R.; Birkel, C. Groundwater recharge mechanisms inferred from isoscapes in a complex tropical mountainous region. *Geophys. Res. Lett.* **2016**, *4343*, 5060–5069. [CrossRef]

18. Crawford, J.; Hughes, C.E.; Parkes, S.D. Is the isotopic composition of event-based precipitation driven by moisture source or synoptic scale weather in the Sydney Basin, Australia? *J. Hydrol.* **2013**, *507*, 213–226. [CrossRef]

19. Jouzel, J.; Delaygue, G.; Landais, A.; Masson-Delmotte, V.; Risi, C.; Vimeux, F. Water isotopes as tools to document oceanic sources of precipitation. *Water Resour. Res.* **2013**, *49*, 7469–7486. [CrossRef]

20. Paar, D.; Mance, D.; Stroj, A.; Pavić, M. Northern Velebit (Croatia) karst hydrological system: Results of a preliminary ^{2}H and ^{18}O stable isotope study. *Geol. Croat.* **2019**, *72*, 205–213. [CrossRef]

21. Kapelj, S.; Kapelj, J.; Švonja, M. Hydrogeological characteristics of the Jadro and Žrnovnica catchment. *Tusculum* **2011**, *5*, 205–216.

22. Biondić, B.; Biondić, R.; Kapelj, S. Karst groundwater protection in the Kupa River catchment area and sustainable development. *Environ. Geol.* **2006**, *49*, 828–839. [CrossRef]

23. Mandić, M.; Bojić, D.; Roller-Lutz, Z.; Lutz, H.O.; Krajcar Bronić, I. Note on the spring region of Gacka River (Croatia). *Isot. Environ. Health Stud.* **2008**, *44*, 201–208. [CrossRef]

24. Frančišković-Bilinski, S.; Cuculić, V.; Bilinski, H.; Häusler, H.; Stadler, P. Geochemical and stable isotopic variability within two rivers rising under the same mountain but belonging to two distant watersheds. *Chem. Erde–Geochem.* **2013**, *73*, 293–308. [CrossRef]

25. Mance, D.; Hunjak, T.; Lenac, D.; Rubinić, J.; Roller-Lutz, Z. Stable isotope analysis of the karst hydrological systems in the Bay of Kvarner (Croatia). *Appl. Rad. Isot.* **2014**, *90*, 23–34. [CrossRef]

26. Ozyurt, N.N.; Lutz, O.H.; Hunjak, T.; Mance, D.; Roller-Lutz, Z. Characterization of the Gacka River basin karst aquifer (Croatia): Hydrochemistry, stable isotopes and tritium -based mean residence time. *Sci. Total Environ.* **2014**, *487*, 245–254. [CrossRef] [PubMed]

27. Horvatinčić, N.; Barešić, J.; Krajcar Bronić, I.; Karman, K.; Forisz, I.; Obelić, B. Study of the bank filtered groundwater system of the Sava River at Zagreb (Croatia) using isotope analyses. *Cent. Eur. Geol.* **2011**, *54*, 121–127. [CrossRef]

28. Marković, T.; Brkić, Ž.; Larva, O. Using hydrochemical data and modelling to enhance the knowledge of groundwater flow and quality in an alluvial aquifer of Zagreb, Croatia. *Sci. Total Environ.* **2013**, *458–460*, 508–516. [CrossRef] [PubMed]

29. Brkić, Ž.; Briški, M.; Marković, T. Use of hydrochemistry and isotopes for improving the knowledge of groundwater flow in a semiconfined aquifer system of the Eastern Slavonia (Croatia). *Catena* **2016**, *142*, 153–165. [CrossRef]

30. Parlov, J.; Kovač, Z.; Nakić, Z.; Barešić, J. Using water stable isotopes for identifying groundwater recharge sources of the unconfined alluvial Zagreb Aquifer (Croatia). *Water* **2019**, *11*, 2177. [CrossRef]

31. Marković, T.; Karlović, I.; Perčec Tadić, M.; Larva, O. Application of stable water isotopes to improve conceptual model of alluvial aquifer in the Varaždin Area. *Water* **2020**, *12*, 379. [CrossRef]

32. Mezga, K.; Urbanc, J.; Cerar, S. The isotope altitude effect reflected in groundwater: A case study from Slovenia. *Isot. Environ. Health. S* **2014**, *50*, 33–51. [CrossRef]

33. Nikolov, J.; Krajcar Bronić, I.; Todorović, N.; Barešić, J.; Petrović Pantić, T.; Marković, T.; Bikit-Schroeder, K.; Stojković, I.; Tomić, M. A survey of isotopic composition (^{2}H, ^{3}H, ^{18}O) of groundwater from Vojvodina. *J. Radioanal. Nucl. Chem.* **2019**, *320*, 385–394. [CrossRef]

34. Lucas, L.L.; Unterweger, M.P. Comprehensive review and critical evaluation of the half-life of tritium. *J. Res. Natl. Inst. Stan.* **2000**, *105*, 541. [CrossRef]

35. Ingraham, N.L. Isotopic variations in precipitation. In *Isotope Tracers in Catchment Hydrology*; Kendall, C., McDonnell, J.J., Eds.; Elsevier: London, UK, 1998; pp. 87–118.

36. Gat, J.R.; Mook, W.G.; Meijer, H.A.J. Atmospheric water. In *Environmental Isotopes in the Hydrological Cycle, Principles and Applications, IHP-V, Technical Documents in Hydrology*; Mook, W.G., Ed.; UNESCO: Paris, France, 2001; pp. 63–74.

37. Rózanski, K.; Gonfiantinu, R.; Araguás-Araguás, L. Tritium in the global atmosphere: Distribution patterns and recent trends. *J. Phys. G Nucl. Part Phys.* **1991**, *17*, 532–536. [CrossRef]

38. Tacher, L.L. The distribution of tritium fallout in precipitation over North America. *Int. Assoc. Hydrol. Bull.* **1962**, *77*, 48. [CrossRef]

39. International Atomic Energy Agency (IAEA). *Environmental Isotopes in the Hydrological Cycle: Principles and Applications*; Gat, J.R., Mook, W.G., Meijer, H.A.J., Eds.; IAEA: Vienna, Austria, 2001; Volume 2, p. 235.

40. Dulinski, M.; Rozanski, K.; Pierchala, A.; Gorczyca, Z.; Marzec, M. Isotopic composition of precipitation in Poland: A 44-year record. *Acta Geophys.* **2019**, *67*, 1637–1648. [CrossRef]

41. Krajcar Bronić, I.; Barešić, J.; Borković, D.; Sironić, A.; Lovrenčić Mikelić, I.; Vreča, P. Long-term isotope records of precipitation in Zagreb, Croatia. *Water* **2020**, *12*, 226. [CrossRef]

42. Barešić, J.; Štrok, M.; Svetek, B.; Vreča, P.; Krajcar Bronić, I. Activity concentration of tritium (^{3}H) in precipitation—Long-term investigations performed in Croatia and Slovenia. In Proceedings of the Sixth International Conference on Radiation and Applications in Various Fields of Research, Ohrid, North Macedonia, 18–22 June 2018; p. 186.

43. Horvatinčić, N.; Kapelj, S.; Sironić, A.; Krajcar Bronić, I.; Kapelj, J.; Marković, T. Investigation of water resources and water protection in the karst area of Croatia using isotopic and geochemical analyses. In Proceedings of the Advances in Isotope Hydrology and its Role in Sustainable Water Resources Management (IHS-2007), Vienna, Austria, 21–25 May 2007; Volume 2, pp. 295–304.

44. Krajcar Bronić, I. Environmental ^{14}C and ^{3}H levels in Croatia. In Proceeding of the 4th International Conference on Environmental Radioactivity: Radionuclides as Tracers of Environmental Processes (ENVIRA 2017), Vilnius, Lithuania, 29 May–2 June 2017; p. 152.

45. Vreča, P.; Krajcar Bronić, I.; Horvatinčić, N.; Barešić, J. Isotopic characteristics of precipitation in Slovenia and Croatia: Comparison of continental and maritime stations. *J. Hydrol.* **2006**, *330*, 457–469.

46. Vreča, P.; Krajcar Bronić, I.; Leis, A.; Brenčič, M. Isotopic composition of precipitation in Ljubljana (Slovenia). *Geologija* **2008**, *51*, 169–180. [CrossRef]

47. Vreča, P.; Krajcar Bronić, I.; Leis, A.; Demšar, M. Isotopic composition of precipitation at the station Ljubljana (Reaktor), Slovenia—Period 2007–2010. *Geologija* **2014**, *57*, 217–230. [CrossRef]

48. Hebert, D. Technogenic tritium in central European precipitations. *Isotopenpraxis* **1990**, *26*, 592–595. [CrossRef]

49. Kazemi, G.A.; Lehr, J.H.; Perrochet, P. *Groundwater Age*; Wiley and Sons: London, UK, 2006; p. 325.

50. Newman, B.D.; Osenbrück, K.; Aeschbach-Hertig, W.; Solomon, D.K.; Cook, P.; Rózanski, K.; Kipfer, R. Dating of "young" groundwaters using environmental trasers: Advantages, applications, and research needs. *Isot. Environ. Health Stud.* **2010**, *46*, 259–278. [CrossRef]

51. Urumović, K.; Hlevnjak, B. The development of basin multilayer aquifer. In Proceedings of the XIX Conference of the Danube Countries on Hydrological Forecasting and Hydrological Bases of Water Management, Osijek, Zagreb, 15–19 June 1998; pp. 519–529.

52. Brkić, Ž. The relationship of the geological framework to the Quaternary aquifer system in the Sava River valley (Croatia). *Geol. Croat.* **2017**, *70*, 201–213. Available online: http://www.geologia-croatica.hr/ojs/index.php/GC/article/view/gc.2017.12/pdf (accessed on 10 July 2020). [CrossRef]

53. Biondić, B.; Biondić, R. Hidrogeologija dinarskog krša u Hrvatskoj. In *Hydrogeology of the Dinaric Karst in Croatia*; Mesec, J., Ed.; University of Zagreb, Faculty of Geotehnical Engineering: Varaždin, Croatia, 2014; p. 341. ISBN 978-953-96597-7-4.

54. Bonacci, O. *Karst Hydrology*; Springer: Berlin/Heilderbelg, Germany, 1987; p. 173.
55. Slišković, I. Vode u kršu slivova Neretve i Cetine. In *Water in the Karst of the Neretva and Cetina River Basins*; Halamić, J., Ed.; Department of Hydrogeology and Engineering Geology: Zagreb, Croatia, 2014.
56. Zaninović, K.; Gajić-Čapka, M.; Perčec Tadić, M.; Vučetić, M.; Milković, J.; Bajić, A.; Cindrić, K.; Cvitan, L.; Katušin, Z.; Kaučić, D.; et al. *Klimatski Atlas Hrvatske/Climate Atlas of Croatia 1961–1990, 1971–2000*; State Hydrometeorological Institute: Zagreb, Croatia, 2008. Available online: https://klima.hr/razno/publikacije/klimatski_atlas_hrvatske.pdf (accessed on 10 July 2020).
57. International Atomic Energy Agency (IAEA). *Stable Isotope Hydrology: Deuterium and Oxygen-18 in Water Cycle*; Gat, J.R., Gonfiantini, R., Eds.; Technical Reports Series 210; IAEA: Vienna, Austria, 1981; p. 339.
58. Coplen, T.B. New guidelines for the reporting of stable hydrogen, carbon, and oxygen isotope ratio data. *Geochim. Cosmochim. Acta* **1996**, *60*, 3359. [CrossRef]
59. Dunn, P.J.H.; Carter, J.F. *Good Practice Guide for Isotope Ratio Mass Spectrometry*, 2nd ed.; FIRMS: Bristol, UK, 2018; p. 84. ISBN 978-0-948926-33-4. Available online: http://www.forensic-isotopes.org/gpg.html (accessed on 10 July 2020).
60. Horvatinčić, N. Radiocarbon and tritium measurements in water samples and application of isotopic analyses in hydrology. *Fizika* **1980**, *12*, 201–218.
61. Krajcar Bronić, I.; Obelić, B.; Srdoč, D. The simultaneous measurement of tritium activity and background count rate in a proportional counter by the Povinec method: Three years' experience at the Rudjer Bošković Institute. *Nucl. Instrum. Methods Phys. Res. B* **1986**, *17*, 498–500. [CrossRef]
62. Barešić, J.; Horvatinčić, N.; Krajcar Bronić, I.; Obelić, B. Comparison of two techniques for low-level tritium measurement—Gas proportional and liquid scintillation counting. In Proceedings of the Third European IRPA Congress, Helsinki, Finland, 14–18 June 2010; pp. 1988–1992.
63. Barešić, J.; Krajcar Bronić, I.; Horvatinčić, N.; Obelić, B.; Sironić, A.; Kožar-Logar, J. Tritium activity measurement of water samples using liquid scintillation counter and electrolytical enrichment. In Proceedings of the 8th Symposium of the Croatian Radiation Protection Association, Krakow, Croatia, 13–15 April 2011; pp. 461–467.
64. Krajcar Bronić, I.; Barešić, J.; Sironić, A.; Horvatinčić, N. Stability analysis of systems for preparation and measurement of 3H and 14C (Analiza stabilnosti sustava za pripremu i mjerenje 3H i 14C, in Croatian with English Abstract). In Proceedings of the 9th Symposium of the Croatian Radiation Protection Association, Krakow, Croatia, 10–12 April 2013; pp. 495–501.
65. Stojković, I.; Todorović, N.; Nikolov, J.; Krajcar Bronić, I.; Barešić, J.; Kozmidic Luburić, U. Methodology of tritium determination in aqueous samples by Liquid Scintillation Counting techniques. In *Tritium—Advances in Research and Applications*; Janković, M.M., Ed.; NOVA Science Publishers: New York, NY, USA, 2018; pp. 99–156.
66. Lobby Factors. Croatian Geological Survey: HGI-CGS Database. Available online: https://lobbyfacts.eu/representative/c1243fabb99641be8382bf2a80b68d61/croatian-geological-survey (accessed on 10 July 2020).
67. Brkić, Ž.; Marković, T. *Jasenik—Additional Hydrogeological Investigations: Technical Report*; Croatian Geological Survey: Zagreb, Croatia, 2014. (In Croatian)
68. Brkić, Ž.; Marković, T.; Larva, O. *A Hydrogeological Study for the Purpose of Defining Groundwater Exploitation Resources in the Area of Koprivnica—Đurđevac: Technical Report*; Croatian Geological Survey: Zagreb, Croatia, 2014. (In Croatian)
69. Larva, O.; Marković, T.; Mraz, V. Hydrodynamic and hydrochemical conditions at the groundwater source "Pašino vrelo", with a focus on its development. *Geol. Croat.* **2010**, *6363*, 299–312. [CrossRef]
70. IGH-IGI. *Geophysical and Hydrogeological Investigations, Dubočanka Valley—Papuk: Technical Report*; Croatian Geological Survey: Zagreb, Croatia, 2003. (In Croatian)
71. Kuhta, M. *Tisovac Spring—Water Supply Požeština (Papuk Nature Park)–Hydrogeological Studies: Technical Report*; Croatian Geological Survey: Zagreb, Croatia, 2014. (In Croatian)
72. Larva, O.; Marković, T.; Mraz, V. *Krapina-Zagorje County—Lobor Pumping Station—Hydrogeological Investigations to Determine Exploitation Groundwater Resources: Technical Report*; Croatian Geological Survey: Zagreb, Croatia, 2014. (In Croatian)
73. Biondić, B.; Kapelj, S.; Kuhta, K. *Water Management Plan of the Republic of Croatia. GIS of Istria: Hydrogeology: Technical Report*; Croatian Geological Survey: Zagreb, Croatia, 1999. (In Croatian)

74. Biondić, B.; Kapelj, S.; Rubinić, J. *Transboundary Aquifers of Croatia and Slovenia between the Kvarner Bay and the Trieste Bay: Technical Report*; Croatian Geological Survey: Zagreb, Croatia, 2004. (In Croatian)

75. Biondić, R.; Biondić, B.; Kapelj, S. *Hydrogeological Investigation of the Novljanska Žrnovnica spring: Technical Report*; Croatian Geological Survey: Zagreb, Croatia, 1997. (In Croatian)

76. Biondić, B.; Kapelj, S.; Biondić, D.; Biondić, R.; Novosel, A. *A Vulnerability Study of the Upper Part Kupa River Basin: Technical Report*; Croatian Geological Survey: Zagreb, Croatia, 2002. (In Croatian)

77. Pavičić, A.; Prelogović, E.; Kapelj, J.; Biondić, D.; Kapelj, S.; Hinić, V. *A Vulnerability Study of the Gacka River. Technical Report*; Croatian Geological Survey: Zagreb, Croatia, 1997. (In Croatian)

78. Pavičić, A.; Terzić, J.; Marković, T. *Hydrogeological Investigation of the in the Wider Area of the Korenica Spring (Phase II): Technical Report*; Croatian Geological Survey: Zagreb, Croatia, 2006. (In Croatian)

79. Pavičić, A.; Terzić, J.; Marković, T.; Dukarić, A. *Hydrogeological Report on the Establishment of the Lička Jasenica Regional Pumping Site: Technical Report*; Croatian Geological Survey: Zagreb, Croatia, 2008. (In Croatian)

80. Kuhta, M.; Božičević, S.; Kapelj, S.; Miko, S. *A Study of the Protection and Use of Valuable Natural Features of the Modrič Cave and Its Surroundings: Technical Report*; Croatian Geological Survey: Zagreb, Croatia, 1999. (In Croatian)

81. Kapelj, S.; Kapelj, S.; Singer, D. *Hydrogeological Investigation of the Jadro and Žrnovnica Catchment Areas (Phase II). Technical Report*; Croatian Geological Survey: Zagreb, Croatia, 2002. (In Croatian)

82. Buljan, R. *Spring Lopuško vrelo—Proposal of the Sanitary Protection Zones: Technical Report*; Croatian Geological Survey: Zagreb, Croatia, 2002. (In Croatian)

83. Dansgaard, W. Stable isotopes in precipitation. *Tellus* **1964**, *16*, 436–468. [CrossRef]

84. Craig, H. Standard for reporting concentrations of deuterium and oxygen-18 in natural waters. *Science* **1961**, *133*, 1833–1834. [CrossRef] [PubMed]

85. Dansgaard, W. The isotope composition of natural waters. *Medd. Grønland* **1961**, *165*, 120.

86. Froehlich, K.; Gibson, J.; Aggarwal, P. Deuterium excess in precipitation and its climatological significance. In Proceedings of the Study of Environmental Change Using Isotope Techniques, Vienna, Austria, 23–27 April 2001; pp. 54–66.

87. Crawford, J.; Hughes, C.E.; Lykoudis, S. Alternative least squares methods for determining the meteoric water line, demonstrated using GNIP data. *J. Hydrol.* **2014**, *519*, 2331–2340. [CrossRef]

88. Hughes, C.E.; Crawford, J. A new precipitation weighted method for determining the meteoric water line for hydrological applications demonstrated using Australian and global GNIP data. *J. Hydrol.* **2012**, *464*, 344–351. [CrossRef]

89. Hunjak, T. Spatial Distribution of Oxygen and Hydrogen Stable Isotopes from Precipitation in Croatia. Ph.D. Thesis, University of Zagreb, Zagreb, Croatia, 2015; 82p. (In Croatian).

90. Hunjak, T.; Lutz, H.O.; Roller-Lutz, Z. Stable isotope composition of the meteoric precipitation in Croatia. *Isot. Environ. Health. S* **2013**, *4949*, 36–345. [CrossRef]

91. Vreča, P.; Krajcar Bronić, I.; Leis, A. Isotopic composition of precipitation in Portorož (Slovenia). *Geologija* **2011**, *5454*, 129–138. [CrossRef]

92. Vreča, P.; Krajcar Bronić, I.; Leis, A. Isotopic composition of precipitation at the station Portorož, Slovenia—Period 2007–2010. *Geologija* **2015**, *5858*, 233–246. [CrossRef]

93. Hunjak, T.; Roller-Lutz, Z. *Interpretation of the Composition of Stable Isotopes of Precipitation in Northern Croatia: Report*; SILab Stable Isotope Laboratory at the Physics Department, School of Medicine, University of Rijeka: Rijeka, Croatia, 2013. (In Croatian)

94. Pfahl, S.; Sodremann, H. What controls deuterium excess in global precipitation? *Clim. Past* **2014**, *10*, 771–781. [CrossRef]

95. Peng, H.; Mayer, B.; Harris, S.; Krouse, H.R. The influence of below-cloud secondary effects on the stable isotope composition of hydrogen and oxygen in precipitation at Calgary, Alberta, Canada. *Tellus* **2007**, *5959*, 698–704. [CrossRef]

96. Vogel, J.C.; Lerman, J.C.; Mook, W.G. Natural isotopes in surface and groundwater from Argentina. *Hydrol. Sci. Bull.* **1975**, *2020*, 203–221.

97. Bottyán, E.; Czuppon, G.; Weidinger, T.; Haszpra, L.; Kármán, K. Moisture source diagnostics and isotope characteristics for precipitation in east Hungary: Implications for their relationship. *Hydrolog. Sci. J.* **2017**, *6262*, 2049–2060. [CrossRef]

98. Posavec, K. Identification and Prediction of Minimum Ground Water Levels of Zagreb Alluvial Aquifer Using Recession Curve Models. Ph.D. Thesis, University of Zagreb, Zagreb, Croatia, 2016; 89p. (In Croatian).
99. Brkić, Ž.; Larva, O.; Urumović, K. Quantity status of groundwater in alluvial aquifers in Northern Croatia. *Geol. Croat.* **2010**, *63*, 283–298. [CrossRef]
100. Barešić, J.; Parlov, J.; Kovač, Z.; Sironić, A. Use of nuclear power plant released tritium as a groundwater tracer. *Rud. Geol. Naft. Zb.* **2019**, *3535*, 25–35. Available online: https://hrcak.srce.hr/ojs/index.php/rgn/article/view/9589 (accessed on 10 July 2020). [CrossRef]

Article

Isotope Composition of Precipitation, Groundwater, and Surface and Lake Waters from the Plitvice Lakes, Croatia

Ines Krajcar Bronić [1],*, Jadranka Barešić [1], Andreja Sironić [1], Ivanka Lovrenčić Mikelić [1], Damir Borković [1], Nada Horvatinčić [1] and Zoran Kovač [2]

[1] Laboratory for Low-level Radioactivities, Division of Experimental Physics, Ruđer Bošković Institute, 10000 Zagreb, Croatia; jbaresic@irb.hr (J.B.); asironic@irb.hr (A.S.); ivanka.lovrencic@irb.hr (I.L.M.); damir.borkovic@irb.hr (D.B.); horvatin@irb.hr (N.H.)

[2] Faculty of Mining, Geology and Petroleum Engineering, University of Zagreb, 10000 Zagreb, Croatia; zoran.kovac@rgn.hr

* Correspondence: krajcar@irb.hr

Received: 5 August 2020; Accepted: 25 August 2020; Published: 28 August 2020

Abstract: The application of tritium, ^{2}H, and ^{18}O in the characterization of the precipitation, groundwater, and surface and lake water of the Plitvice Lakes (PL), Croatia, over the 1979–2019 period is presented. An increase in the mean annual air temperature of 0.06 °C/year and in the annual precipitation amount of 10 mm/year is observed. The good correlation of the tritium activity concentration in the PL and Zagreb precipitation implies that the tritium data for Zagreb are applicable for the study of the PL area. The best local meteoric water line at PL was obtained by the reduced major axis regression (RMA) and precipitation-weighted ordinary least squares regression (PWLSR) approaches: $\delta^2H_{PWLSR} = (7.97 \pm 0.12)\,\delta^{18}O + (13.8 \pm 1.3)$. The higher deuterium excess at PL (14.0 ± 2.2 ‰) than that at Zagreb reflects the higher altitude and influence of the Mediterranean precipitation. The δ^2H in precipitation ranges from −132.4‰ to −22.3‰ and δ^{18}O from −18.3 ‰ to −4.1‰. The much narrower ranges in the groundwater (<1‰ in δ^{18}O, <10‰ in δ^2H) indicate the good mixing of waters in aquifers and short mean residence times. The higher average δ^2H in all three karst springs observed after 2003 can be attributed to the increase in the mean air temperature. The mean δ^2H and δ^{18}O values in the surface and lake water increase downstream due to the evaporation of surface waters. There is no significant difference between the surface water line and the lake water line (2011–2014). The stable isotope composition of the surface and lake waters reacts to extreme hydrological conditions.

Keywords: Plitvice Lakes; Croatia; karst; precipitation; groundwater; surface water; lake water; tritium; deuterium δ^2H; oxygen δ^{18}O

---

1. Introduction

Karst is a special type of landscape that is formed by the process of karstification—i.e., the dissolution of soluble carbonate rocks, mostly limestone and dolomite. Karst channels, conduits, and fissures store relatively large quantities of groundwater, and such karst aquifers are capable of providing large supplies of water for human consumption. The precipitation water quickly infiltrates underground, creating a system of interconnected flow paths, and eventually re-appears at the surface as springs. Karst aquifers, especially in areas with a high permeability, can be very vulnerable to contamination and can enable the fast transport of contaminants through the aquifers, which can result in the degradation of water quality [1–4]. The assessment of the impact of human activities and recent climate changes on karst waters has to be properly considered [5]. A good understanding of the characteristics of karst aquifers, especially the origin of karst spring water, is essential for their efficient protection.

The most prominent methods for the research of the origin of spring water (groundwater) are isotopic methods, especially the application of ^{2}H, ^{18}O, and ^{3}H isotopes that constitute water molecules. They behave as conservative ideal tracers, with a broad application in hydrogeology [5–18]. Precipitation presents an input to groundwater, and therefore knowledge of the isotope composition of precipitation is a prerequisite for groundwater studies [19]. The importance of water isotopes as perfect tracers was recognized by the World Meteorological Organization (WMO) and the International Atomic Energy Agency (IAEA), which established a worldwide network for the monitoring of water isotopes—the Global Network of Isotopes in Precipitation (GNIP) [20–22]. GNIP contains data from Zagreb, as a permanent GNIP station, and from some other locations where the isotope composition of precipitation was monitored in different projects, including the Plitvice Lakes station [23] and references therein. Some differences in the isotopic composition of precipitation in the continental and maritime stations were discussed [19], and the long-term data for the station in Zagreb showed the influence of climate change on the isotope composition [23].

Croatia is a southeastern European country situated between the eastern Adriatic coast and the Pannonian Plain (Figure 1). Dinaric karst covers almost half of the Croatian territory [24,25], including the islands and the Adriatic coast, the high mountain regions, and part of central Croatia. Dinaric karst, known worldwide as the *locus typicus* of classical karst, is a part of the Dinarides system and consists of very porous and permeable rocks, with many permanent and intermittent springs and a developed underground drainage system [26,27]. Croatia is rich with clean drinking water, and in many cases this water is groundwater originated from the karst springs [11,12,14,28].

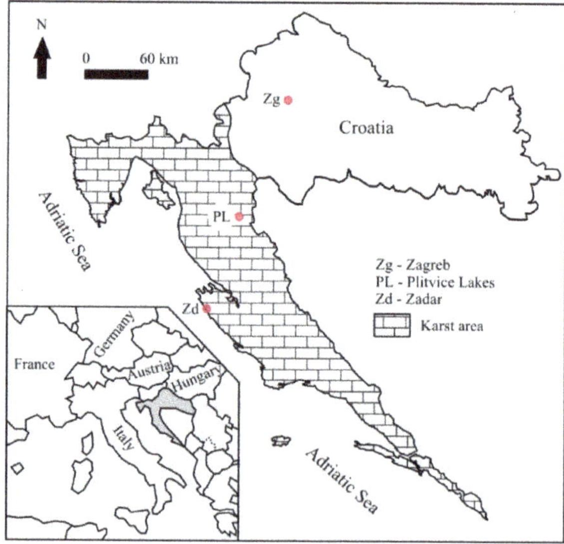

Figure 1. Map of Croatia and the position of the study area: Zagreb (Zg), Plitvice Lakes (PL), and Zadar (Zd).

The Plitvice Lakes (PL) area presents a unique system of 16 cascade flow-through lakes that are fed by three main springs (Crna Rijeka River, Bijela Rijeka River, Plitvica River) and outflow to the Korana River. It is famous worldwide for its beauty and diversity. The area has been protected as a part of the national park since 1949, and since 1979 it has been included in the United Nations Educational, Scientific, and Cultural Organisation (UNESCO) World Heritage List. The area is one of the most studied karst areas in Croatia, where various scientific studies have been performed since the beginning of 20th century. During the 1970s, a group from the Ruđer Bošković Institute joined the scientific community with the application of isotope methods. A comprehensive overview of the investigation conducted at the Plitvice Lakes is given in [29]. An overview of the isotope investigations in both the karst and alluvial aquifers in Croatia [30] at approximately 100 sites in period 1997–2014 presents an important contribution to regional knowledge of groundwater hydrology. However, the study did not include the area of the Plitvice Lakes.

The main aim of the investigation of the group from the Ruđer Bošković Institute was the study of secondary carbonate precipitation in the form of tufa and lake sediment (e.g., [7,31,32]), which made the Plitvice Lakes a unique natural phenomenon. The analysis of more than 30 years of records of various physico-chemical parameters presented geochemical conditions for tufa precipitation in relation to climate change [33]. Isotope methods were applied also to studies of springs and surface and lake waters of the area [7,8,34,35]. Precipitation was also occasionally studied [36,37]. Both the isotope and physico-chemical data at springs showed constant values in different seasons, implying that the water was of atmospheric (meteoric) origin, of both winter and summer precipitation, and that the recharge water was well mixed with the existing water in aquifers [7]. The short mean residence time (MRT) of the water was determined based on the tritium activity concentration [7,8,35] and by the stable isotopes and additional tracers (helium and neon, chlorofluorocarbons (CFCs), and sulfur hexafluoride) [10].

Plitvice Lakes, although protected from direct anthropogenic influence, cannot be protected from global changes such as climate change. Thus, the consideration of the lakes' hydrology is of most importance for the system, especially in dry summer periods [38–40]. Water warming was observed in the surface water and springs of the Crna Rijeka and Bijela Rijeka Rivers [33]. The warming of the waters did not endanger the tufa precipitation process, since the increase in temperature favors both the physico-chemical and biological factors of authigenic calcite precipitation. However, an increase in temperature contributes to the loss of water from the lakes through evaporation [10,38,40].

The aim of this paper is to present various isotope studies of different types of water bodies (precipitation, groundwater, surface lake and river water, lake water from traps at certain depths) from the early period of isotope applications (since 1979) to the most recent one (2018) at the Plitvice Lakes National Park, as an area of unique geomorphological karst formation. Although the Plitvice Lakes are of great scientific interest worldwide, an overview of the isotopic studies of its water bodies has never been presented so far.

The results will be compared to the long-term isotope-in-precipitation data at Zagreb [23], as well as to regional precipitation and groundwater data [30]. The aim of the paper is to evaluate the most important hydrological inputs to the Plitvice Lakes, detect the possible influence of climate change on karst groundwater, and eventually show what conclusions could be drawn after a long-term and rather comprehensive study of a certain area by isotopic techniques.

2. Materials and Methods

2.1. Site Description

The Plitvice Lakes area is located in a continental, mountainous part of west Croatia (Figure 1). The Plitvice Lakes are a system of 16 lakes developed on carbonate rocks [41] separated by tufa barriers and interconnected by waterfalls (Figure 2). Geologically, they belong to the External Dinarides or Dinaric-coastal area, with characteristic carbonate sediments and karstic features. The altitude difference of the system from the springs to the Korana River is about 300 m (Figure 2b). However, the recharge area extends to higher mountains (the highest altitude of the national park is 1279 m a.s.l.), and the average recharge altitude is about 900 m a.s.l. [38].

Table 1. List of sampling locations, their codes, and the types of water body samples.

Location	Code	Sample Type	Comment
Plitvice Lakes	PL	precipitation	study area
Zagreb	Zg	precipitation	comparison location
Zadar	Zd	precipitation	comparison location
Crna Rijeka River	CR	groundwater	spring
Bijela Rijeka River	BR	groundwater	spring
Plitvica River	PR	groundwater	spring
Matica	Ma	surface water	
Lake Prošćansko	LP	surface water	
Lake Ciginovac	LC	surface water	
Lake Burget	LB	surface water	
Burget—Waterfall	BW	surface water	
Lake Kozjak—Bridges	KzB	surface water	
Korana River—Sastavci	KoS	surface water	
Korana River—Bridge	KoB	surface water	
Lake Kozjak	IRB1	lake water	water depth 6 m
Lake Kozjak	IRB2	lake water	water depth 8–10 m
Lake Prošćansko	IRB3	lake water	water depth 6 m
Lake Gradinsko	IRB4	lake water	water depth 2 m

Zagreb is also situated in a continental part of Croatia (to the north) (Figure 1). It belongs to the Pannonian area characterized by its milder relief; predominantly magmatic, clastic, and metamorphic rocks; and well-developed stream grid [42]. Zadar is located at the Adriatic Sea coast (southern Croatia) (Figure 1), which belongs to the Dinaric coastal area.

The three selected locations of precipitation sampling differ in the altitude of the meteorological stations; Plitvice Lakes station is situated at 550 m a.s.l., Zagreb-Grič station at 165 m a.s.l., and Zadar at 5 m a.s.l. [19,23]. Climatologically, Zagreb and Plitvice Lakes belong to the Cfb climate class, characterized by a temperate climate without a dry season and with a warm summer, while Zadar belongs to the Cfa climate class, characterized by a temperate climate without a dry season but with a hot summer [43,44]. The annual precipitation amounts in Zagreb (measured at the Zagreb-Grič meteorological station) and Zadar are almost identical: 883 mm and 915 mm for the 1961–1990 period, and 884 mm and 882 mm for the 1991–2004 period, respectively [19]. The annual precipitation amount is significantly higher at the Plitvice Lakes, where it ranges between 1148 and 2113 mm in the 1986–2019 period [45]. The monthly precipitation at the Plitvice Lakes is distributed relatively uniformly throughout the year, with slight maxima observed in spring and autumn and minimum values in summer months. The annual temperatures range from 8.0 to 10.8 °C (1986–2019), with an average value of 9.2 ± 0.5 °C. January is the coldest month (0 ± 2.3 °C on average) and July is the warmest month (18.4 ± 1.1 °C) [45]. Snow falls between November and March; however, reduced snowfall is observed in recent times compared to the older data [33,45].

The water temperature at the springs is very stable throughout the year. Nevertheless, an increase in the spring temperatures is observed between the 1981–1986 and 2010–2014 periods in both Crna Rijeka spring (from 7.80 ± 0.15 °C to 8.04 ± 0.14 °C) and the Bijela Rijeka spring (from 7.46 ± 0.17 °C to 8.14 ± 0.53 °C) [33]. The temperature of the surface water downstream ranges from near zero in winter to 23 °C in summer, being on the average about 12 °C [33,46]. The lakes can be frozen in the winter months.

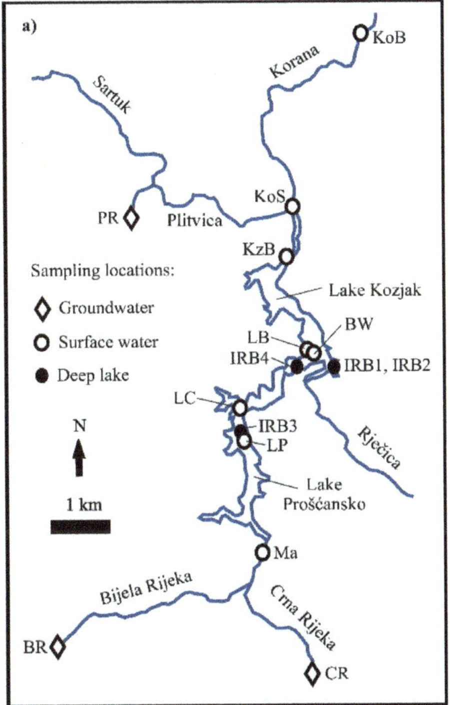

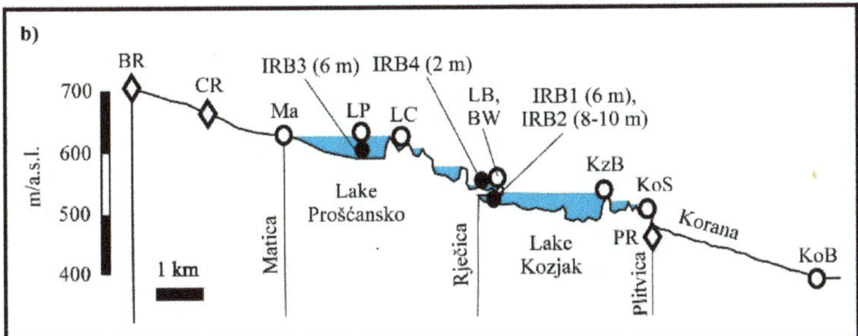

Figure 2. Sampling locations: (**a**) along the water course (top view); (**b**) along the profile. For an explanation of the location codes, see Table 1.

2.2. Sampling

The monthly precipitation was collected between 1978 and 1984 [36] (not continuously) and from 2003 to 2006 [11,37] at the meteorological station at the Plitvice Lakes (altitude 550 m, International Atomic Energy Agency (IAEA) and World Meteorological Organization (WMO) station code 1432501) [23].

Groundwater samples were collected at the three main karst springs of the system: Crna Rijeka River and Bijela Rijeka River (in the south), and the Plitvica River that joins the lake water after the Great waterfall before the location of Korana River—Sastavci (Figure 2, Table 1).

Surface water was collected as grab samples at 8 locations along the water course for the length of ~10 km from Matica (the main stream feeding the lakes) to the Korana River (the outflow from the lake system) (Figure 2, Table 1). Lake water was collected at 4 sediment traps in different lakes:

IRB1 and IRB2 in Lake Kozjak at water depths of 6 m and 8–10 m, respectively; IRB3 in Lake Prošćansko at 6 m; and IRB4 in Lake Gradinsko at 2 m water depths (note: the water from sediment traps will be called "lake water" in the further text). At the same time, the surface water was collected at nearby locations.

All the sampling locations are listed in Table 1, with the location code and the type of water body sampled. The amount of tritium activity concentration data, stable isotope data, and period of sampling is given in Table S1.

2.3. Meteorological and Hydrological Data

The meteorological data consisted of the monthly precipitation amount (P) and the average monthly air temperature (T). The data were obtained on request from the Croatian Meteorological and Hydrological Service (CMHS) [45]. The meteorological records for the Plitvice Lakes, Croatia, exist for three distinctive periods: 1986–1990, 1996–2011, and 2015–2019. The data are not complete for all years, and in further analyses only the years with all 12 months of data will be used. The minimal and maximal monthly values within a year were identified, and the mean annual temperature and total annual precipitation amount were determined. CMHS [45] also provided data on the flow rates measured at different points in the Plitvice Lakes area for the period since 1982, except for the 1991–2001 period.

2.4. Measurement

Details on the measurement techniques and their changes were described in [23], and here we give only a short overview. The stable isotope composition of water samples for the period up to 2003 was measured on a Varian MAT 250 dual inlet isotope ratio mass spectrometer (IRMS) at the Jožef Stefan Institute in Ljubljana [23,47]. The measurement precision of duplicates was better than ± 0.1‰ for $\delta^{18}O$ and ± 1‰ for $\delta^{2}H$. The $\delta^{2}H$ and $\delta^{18}O$ in the surface and lake water samples in the period 2012–2014 were measured at the JOANNEUM, Graz, Austria. The oxygen isotopic composition was determined on a dual-inlet Finnigan DELTAplus by means of the fully automated equilibration technique, and the isotopic composition of hydrogen was determined on a continuous flow Finnigan DELTAplus XP mass spectrometer with a HEKAtech high-temperature oven by the reduction of water over hot chromium [48]. All the measurements were carried out together with laboratory standards that were calibrated periodically against international standards, as recommended by the IAEA. The measurement precision was better than ± 0.1‰ for $\delta^{18}O$ and ± 1‰ for $\delta^{2}H$ [47]. The stable isotope composition of recent samples was determined at the Laboratory for Spectroscopy of the Faculty of Mining, Geology, and Petroleum Engineering, University of Zagreb, with a Liquid Water Isotope Analyzer (LWIA-45-EP, Los Gatos Research), and the official LGR working standards were used. The data were analyzed by the Laboratory Information Management System (LIMS) [49].

The tritium activity concentration (A) in all the samples was determined at the Ruđer Bošković Institute in Zagreb, except for some data for the 2003–2006 period [10]. The results are expressed in tritium units (1 TU = 0.118 Bq l^{-1}) [20,22], which represent one ^{3}H atom in 10^{18} atoms of hydrogen. The gas proportional counting technique (GPC) was used up to 2009 [34,50]. The detection limit was 2.5 TU, and the measurement uncertainty was between 2 and 5 TU, depending on the activity concentration. Since 2008, the technique of the electrolytical enrichment of samples (LSC-EE) and measurement by an ultra-low-level liquid scintillation counter Quantulus 1220 was used [51]. The detection limit obtained by the LSC-EE technique was 0.5 TU, and the measurement uncertainty was between 0.5 and 3 TU [51].

2.5. Data Analysis

The results of the stable isotope analyses are reported as δ-values—i.e., the relative difference in isotope ratios of the sample and the standard [52]:

$$\delta_{S/R} = \frac{R_{Sample}}{R_{Reference}} - 1, \tag{1}$$

where R_{Sample} and $R_{Reference}$ stand for the isotope ratio ($R = {}^2H/{}^1H$ and $R = {}^{18}O/{}^{16}O$) in the sample and the reference material (standard), respectively. The δ-values are dimensionless and small, and therefore they are expressed in per mill (‰) [21,52–54]. The international standard VSMOW (the Vienna Standard Mean Ocean Water) is used [23,54,55]. The δ^2H and $\delta^{18}O$ isotopic compositions of meteoric waters (precipitation and atmospheric water vapor) are strongly correlated, and the relation in Equation (2) is referred to as the global meteoric water line (GMWL) [53,55–57].

$$\delta^2H = 8.0 \cdot \delta^{18}O + 10 \tag{2}$$

The GMWL describes the general relation between δ^2H and $\delta^{18}O$ on a global scale reasonably well. However, for applications in hydrogeological studies, regional local meteoric water lines (LMWLs), either long-term or for certain shorter periods, can be more appropriate [23,58,59]. Generally, a LMWL has the form $\delta^2H = a\, \delta^{18}O + b$, where a is the slope and b is the intercept. LMWLs can differ from the GMWL in terms of both the slope and intercept values, depending on the conditions for forming a local water source [21,58–60].

The deuterium excess (d-excess, or d) was calculated from the paired monthly data according to Equation (3) [56]:

$$d = \delta^2H - 8\, \delta^{18}O. \tag{3}$$

This can be related to the meteorological conditions in the source region from which the water vapor is obtained [20,21,58–60]. Autumn and winter precipitation originating from the Mediterranean Sea is characterized by distinctly higher d-excess values ($d > 18‰$) than precipitation coming from the Atlantic ($d \sim 10‰$), reflecting the specific source conditions during water vapor formation [19,59,60].

Correlations between various data points were obtained as ordinary least squares regressions using standard commercial software. Pearson's coefficient r is given, as are the number of data pairs n and the p-values describing the statistical significance of the correlations. The data taken from the literature usually have the adjacent R^2 value reported.

In the special case of calculating the a- and b-values of LMWLs, different methods were applied: ordinary least squares regression (OLSR), reduced major axis regression (RMA), and major axis least squares regression (MA, or the orthogonal regression) [61,62]. We calculated precipitation-weighted regressions (PWLSR, PWRMA, and PWMA) [61–63], which took into account the precipitation amount in a particular month. The local meteoric water lines are defined as $LMWL_{OLSR}$, $LMWL_{RMA}$, $LMWL_{MA}$, $LMWL_{PWLSR}$, $LMWL_{PWRMA}$, and $LMWL_{PWMA}$. For calculating the regressions from data sets of at least 36 continuous monthly records, we used the Local Meteoric Water Line Freeware [64]. The software also calculated an average of the root mean square sum of squared errors ($rmSSEav$), which is a relative error that allows for a comparison of different methods; the closer the value of $rmSSEav$ is to 1.0, the better the regression method is for that set of data [61].

3. Results and Discussion

In this section, we present an analysis of the climatological parameters (temperature T and precipitation amount P) in the area of the Plitvice Lakes, followed by the isotope composition (tritium activity concentration, δ^2H and $\delta^{18}O$ values) of precipitation, groundwater, and finally the surface and lake waters.

3.1. Temperature and Precipitation Amount

The monthly mean temperatures in three periods when data are available—1986–1990, 1996–2011, and 2015–2019—show an increase in temperature in all months (Figure 3a). The increase is statistically significant at 5% ($p < 0.05$) in June, July, August, and November, and at 10% ($p < 0.10$) in April. When the mean annual temperatures in the 1986–2019 period are compared (Figure 3b) (only years with data for all 12 months), an increase with the slope of 0.06 ± 0.01 °C per year is observed, and $n = 16$, $r = 0.85$, $p < 0.05$. The mean annual temperature at the Zagreb-Grič station in the 1976–2018 period showed a significant increase, with a slope of 0.071 ± 0.008 °C per year ($r = 0.82$, $p < 0.05$) and a faster increase in the maximal annual temperatures (0.09 ± 0.02 °C per year, $r = 0.69$) [23], which is in accordance with the observation at the Plitvice Lakes, although the set of meteorological data at the Plitvice Lakes is not complete. The mean temperatures in the three periods with available data (Table 2) show a constant increase.

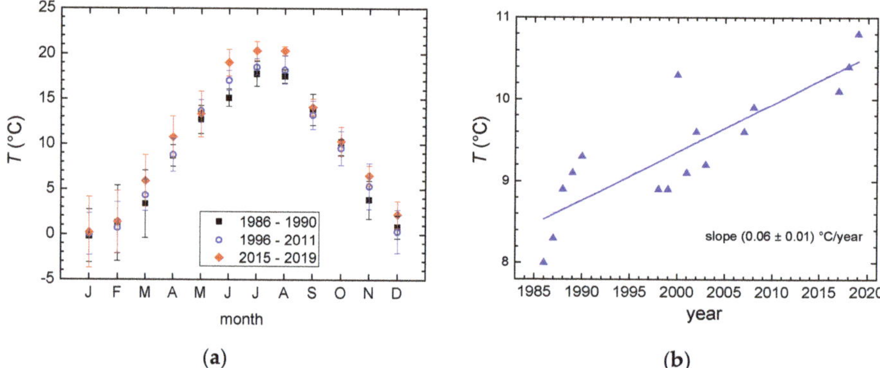

Figure 3. (**a**) Monthly mean temperatures at the Plitvice Lakes meteorological station in the three periods when data are available; (**b**) mean annual temperatures at the Plitvice Lakes in 1986–2019 (only years with complete data for 12 months are shown).

Table 2. Mean annual temperature and annual precipitation amount in the three periods with available meteorological data.

Period	T (°C)	P (mm)
1986–1990	8.7 ± 0.5	1480 ± 170
1996–2011	9.4 ± 0.5	1580 ± 330
2016–2019	10.4 ± 0.4	1745 ± 60

The annual increases in temperature at both the Zagreb and PL stations can be compared with the global temperature increase and the temperature increase in the city of Ljubljana in Slovenia. Ljubljana shows a distinctive air warming trend, particularly in the period 1979–2008, of 0.06 °C per year [65]. Although this period does not include the hottest years on record globally (2014–2018), with 2016 being the hottest year [66], the value of the increase is consistent with the values obtained for Zagreb [23] and the Plitvice Lakes. Moreover, all the data are higher than the globally observed temperature change of 0.018 °C per year in the 1980–2020 period [67]. A recent study [68] indicated that the cities in East Europe, including Ljubljana, will experience air warming at a higher extent than is observed globally.

The monthly mean precipitation amounts show no significant seasonal variations (Figure 4a), although a slight minimum is observed in the summer months, and slightly higher values in spring and autumn. Higher fluctuations within a month are observed in the most recent period (2015–2019) compared to the older ones (Figure 4a). The annual precipitation amounts increase with a slope of

10 ± 7 mm per year, $r = 0.31$ (Figure 4b, Table 2). A wider range in the monthly and annual precipitation amount values has been observed (Figure 4b, Table 2), as observed also in the data for Zagreb [23]. The long-term (1976–2018) annual precipitation amount at Zagreb showed a slight, statistically not significant, increase of 1.4 ± 1.7 mm per year. However, the most prominent characteristics of the data were the higher deviations of the annual values in the period after 2000 from the mean value for the whole period [23].

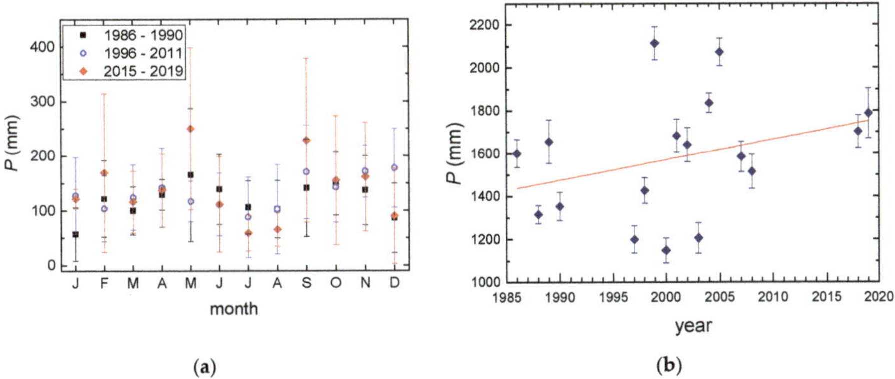

Figure 4. (**a**) Monthly mean precipitation amount at the Plitvice lakes meteorological station in the three periods when data are available; (**b**) annual precipitation amounts at the Plitvice Lakes in 1986–2019 (only years with complete data for 12 months are shown).

3.2. Isotope Composition of Precipitation

The record of the tritium activity concentration (A) in the precipitation for Plitvice Lakes (Table S2) is not as complete as the one for Zagreb precipitation [23] (Figure 5). The most complete data for Plitvice Lakes are recorded from mid-1980 to mid-1982, and later from mid-2003 to mid-2006. Both records exhibited a seasonal pattern typical of the continental stations of the Northern Hemisphere; the maximal monthly ^{3}H activity concentrations were observed between May and July, mostly in June, and the lowest ^{3}H activity concentrations were observed in winter. Seasonal variations were superposed on the basic decreasing trend of the mean annual values. The data for Zagreb show that, after 1996, there is no significant decrease in the A values of precipitation. The mean value of A in the precipitation at Zagreb for the 1995–2018 period was 8.5 ± 1.2 TU [23], and in the 2003–2006 period it was 7.7 + 1.8 TU, while that for the precipitation at Plitvice Lakes in the 2003–2006 period was 7.4 ± 4.5 TU, with the highest value of 18.8 TU in July 2006 and winter values close to the detection limit of the GPC method.

It is obvious (Figure 5) that the tritium activity concentration in the precipitation at Plitvice Lakes follows the trend of the precipitation in Zagreb. The linear correlation between the two sets revealed a line with a slope of (1.02 ± 0.04), $n = 62$, $r = 0.96$, and the paired t-test showed that at the 0.05 level, the two sets of data are not significantly different.

The δ^2H values in the Plitvice Lakes precipitation range from −132.4‰ (February 2005, monthly temperature −4.1 °C) to −22.3‰ in August 2003 (monthly temperature 21.7 °C) (Table S2, Figure 6). The extreme δ^{18}O values, −18.3% and −4.1‰, were observed in the same months. The mean δ^2H and δ^{18}O values in the period from July 2003 to September 2006 are −65.3‰ and −9.9‰, respectively. However, if the 3-year cycle is taken into account (July 2003–June 2006) the corresponding values are −67.5‰ and −10.2‰, respectively. These values can now be compared to the Zagreb values in the same period 2003–2006, which are −59.7‰ and −8.3‰ for δ^2H and δ^{18}O, respectively, resulting in difference of 1.9‰ in δ^{18}O between Zagreb and the Plitvice Lakes. Differences in the mean annual temperatures at Zagreb (12.3 °C [23]) and the Plitvice Lakes (9.4 °C, Table 2) can account for about 1‰ difference in the δ^{18}O values, taking into account the temperature gradient of 0.331‰ per °C [23].

A difference in altitude of 385 m and the altitude effect of 0.28‰ per 100 m altitude difference [19] would account for a further 1.1‰, giving a total difference of 2.1‰ in $\delta^{18}O$, which is in good agreement with the observed difference, taking into account limited number of isotope data for the Plitvice Lakes.

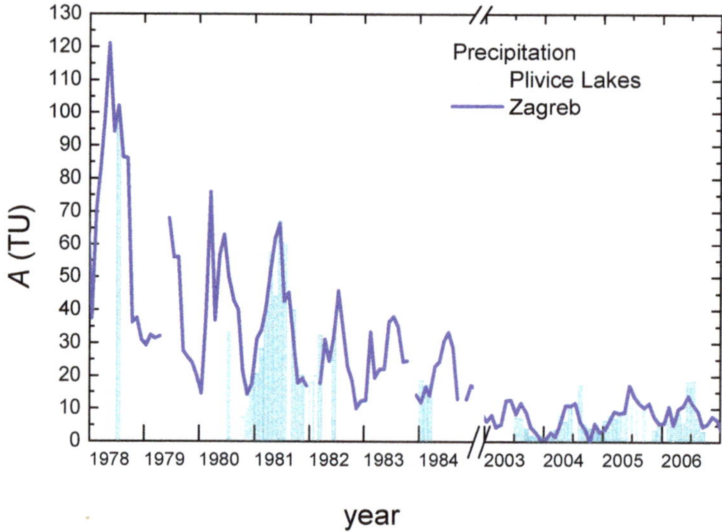

Figure 5. Tritium activity concentration in the precipitation at Zagreb and Plivice Lakes for the 1978–1984 and 2003–2006 periods.

Table 3. Values of slopes ($a \pm \sigma_a$) and intercepts ($b \pm \sigma_b$) for LMWLs obtained by different regression methods for precipitation at the Plitvice Lakes, Zadar, and Zagreb. OLSR: ordinary least squares regression; RMA: reduced major axis regression; MA: major axis least squares regression; PWLSR, PWRMA, and PWMA: precipitation-weighted respective regressions; n—number of data points included; $rmSSEav$—average of the root mean square sum of squared errors [60]. In bold: approaches that describe LMWL the best.

Location and Period	Method	n	$a \pm \sigma_a$	$b \pm \sigma_b$	$rmSSEav$
Plitvice Lakes	OLSR	36	7.85 ± 0.12	12.46 ± 1.23	1.0016
2003–2006	**RMA**	36	**7.88 ± 0.12**	**12.77 ± 1.20**	**1.0006**
	MA	36	7.91 ± 0.12	13.06 ± 1.24	1.0016
	PWLSR	36	**7.97 ± 0.12**	**13.82 ± 1.32**	**1.0000**
	PWRMA	36	8.00 ± 0.12	14.15 ± 1.32	1.0073
	PWMA	36	8.04 ± 0.12	14.47 ± 1.32	1.0162
2003–2005 [10]	OLSR	24	7.85 ± 0.12	12.46 ± 1.23	1.0016
Zadar	OLSR	36	6.57 ± 0.41	1.55 ± 2.37	1.0262
2003–2004	**RMA**	36	**6.70 ± 0.40**	**3.76 ± 2.31**	**1.0100**
Data from [19]	MA	36	7.43 ± 0.440	6.01 ± 2.52	1.0259
	PWLSR	36	**7.27 ± 0.50**	**5.95 ± 3.03**	**1.0000**
	PWRMA	36	7.84 ± 0.51	9.23 ± 3.09	1.0525
	PWMA	36	8.44 ± 0.54	12.67 ± 3.26	1.1433
Zagreb	OLSR	121	7.43 ± 0.11	2.59 ± 0.97	1.0048
1996–2006	**RMA**	121	**7.52 ± 0.10**	**3.35 ± 0.96**	**1.0019**
[23]	MA	121	7.60 ± 0.11	4.08 ± 0.98	1.0047
	PWLSR	121	7.38 ± 0.11	2.68 ± 0.95	1.0095
	PWRMA	121	7.47 ± 0.11	3.44 ± 0.95	1.0024
	PWMA	121	**7.56 ± 0.11**	**4.19 ± 0.96**	**1.0018**
1980–2018 [23]	**RMA**	389	**7.74 ± 0.06**	**5.57 ± 0.55**	**1.0019**
RMWL_cont 2008–2013 [30]	OLSR	524	7.4 ± 0.005	4.1 ± 0.5	R = 0.99
RMWL_coast 2008–2013 [30]	OLSR	655	7.0 ± 0.08	4.4 ± 0.5	R = 0.96

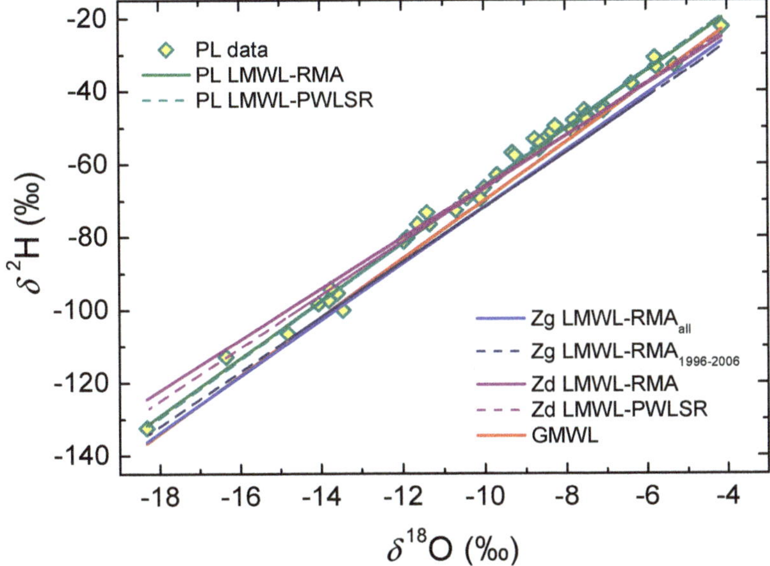

Figure 6. Stable isotope composition of the precipitation at the Plitvice Lakes (PL) and the LMWLs obtained by the reduced major axis regression (RMA) and precipitation-weighted ordinary least squares regression (PWLSR) methods (Table 3). Local Meteoric Water Lines (LMWLs) for Zagreb (RMA method) and for Zadar (RMA and PWLSR methods) as well as the Global Meteoric Water Line (GMWL)(δ^2H = 8 δ^{18}O + 10) are shown for comparison.

In order to determine the relation between the δ^2H and δ^{18}O values in the precipitation of Plitvice Lakes—i.e., the LMWL of the form δ^2H = ($a \pm \sigma_a$) δ^{18}O + ($b \pm \sigma_b$)—we performed linear regression by different approaches (Table 3). For comparison, the analysis for the Zagreb precipitation is taken from [23], while the data for Zadar [19] were analyzed here also by applying the same approaches (Table 3). The analysis of the stable isotope composition of the Plitvice Lakes precipitation is based on 36 data pairs from 2003–2006 (Table 1, Table S2). The LMWLs obtained by various correlation methods for all three precipitation stations (PL, Zg, and Zd) are shown in Table 3. The best results for the Plitvice Lakes and Zadar precipitation were obtained by the RMA and PWLSR approaches (marked in bold in Table 3), while the Zagreb precipitation data were best described by the RMA and PWMA approaches (marked in bold in Table 3). The LMWLs obtained by the best approaches are shown in Figure 6. The LMWL$_{RMA}$ and LMWL$_{PWLSR}$ for PL have slopes close to slope 8 of the Global Meteoric Water Line (GMWL), but slightly higher intercepts. The slope of the Zd LMWLs is lower, as well as the intercepts, caused by evaporation during the summer months [19]. In general, the Plitvice Lakes LMWL lines lie above the GMWL and LMWLs for Zagreb, and are close to the LMWLs for Zadar (Figure 6). Several precipitation sampling campaigns of precipitation related to the cave drip water studies in the karst area of Croatia revealed similar results for LMWLs: sites along the coast at low altitudes gave values of slope (6.6–6.8) and intercept (3.8–6.7) similar to those obtained for Zadar, while continental stations gave slopes (7.1 to 7.8) and intercepts (2.3 to 14.5) closer to the ones of Zagreb and the Plitvice Lakes [25].

The deuterium-excess monthly values at the Plitvice Lakes range from 7.7‰ (January 2004) to 17.9‰ (January 2005), with the mean value of 14.0 ± 2.2‰. These values are higher than those for the Zagreb precipitation, where the mean annual *d*-excess values ranged between 2.3‰ and 10.7‰, and in the 2003–2006 period the mean value was 8.8 ± 0.8‰ [23]. The higher *d* value may be explained by the observed increase in the *d*-excess value with the altitude [2,19], but also by the more intense influence of the Mediterranean air masses in the area of Plitvice lakes [13,19], which is shown here by the relative position of the LMWL compared to the GMWL (Figure 6). Similar high *d*-excess values (between 11.0 and 18.6) were found for several precipitation sampling locations in the continental karst area at altitudes between 300 and 1550 m [25].

The deuterium excess values for the Zagreb precipitation were higher in autumn than in spring. [19,23,69]. The higher *d*-excess in autumn indicates a higher influence of the Mediterranean air masses in these months. A similar pattern of the monthly *d*-excess distribution was observed for the Plitvice Lakes station (Figure 7), with higher *d*-excess monthly values observed in the autumn–winter precipitation (September–December, 15.2 ± 0.4‰). The lowest mean monthly *d*-excess values were observed in the summer (July–August, 12.5 ± 1.0‰).

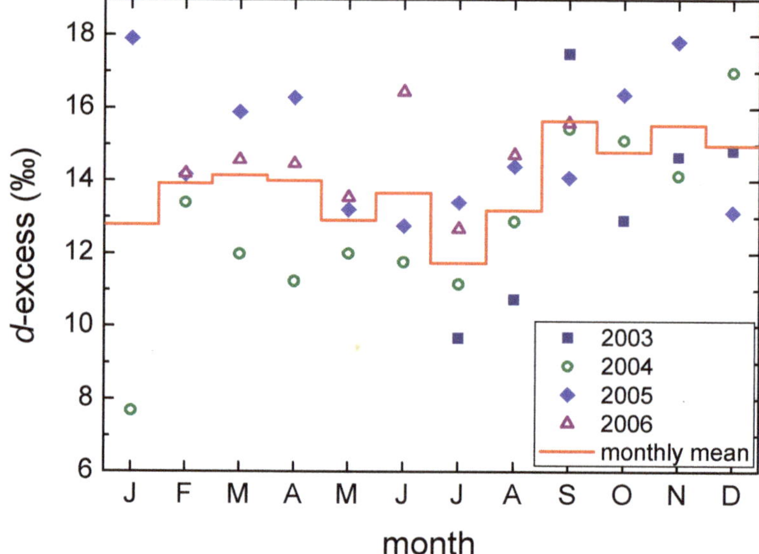

Figure 7. Deuterium-excess values in the precipitation at Plitvice Lakes, 2003–2006 monthly data, and mean monthly values.

It should be mentioned here that the δ^{18}O values from the 1983–1984 period ranged from −7.2‰ to −13.1‰, the δ^2H from −46.0‰ to −94.4‰, and the *d*-excess from 8.9‰ to 11.5‰, based on only five data points for the monthly isotope composition of the precipitation at the Plitvice Lakes [7,8].

3.3. Groundwater

The tritium activity concentration in the three main springs of the Plitvice Lakes (Figure 8) shows a general decrease over the sampling period, but the seasonal variations are much smaller than those for precipitation (Figure 5), and its values fluctuate around the mean annual *A* values for precipitation. The *A* values in the Bijela Rijeka springs were always higher than in the other two springs.

Due to a constant decrease in the *A* values and the non-equal number of data for each spring, we compared the *A* values in the Crna Rijeka and Bijela Rijeka springs in 1984 and 2015 (Figure 9a). The lower values in 2015 are not surprising, as well as the higher *A* values in BR than in CR (by a factor of about 1.3 in 1984 and 1.4 in 2015). In addition, the relative standard deviations of 7.8% in 1984 and 1.3% in 2015 in *A* for BR were smaller in both periods than those for CR (17% and 9%, respectively).

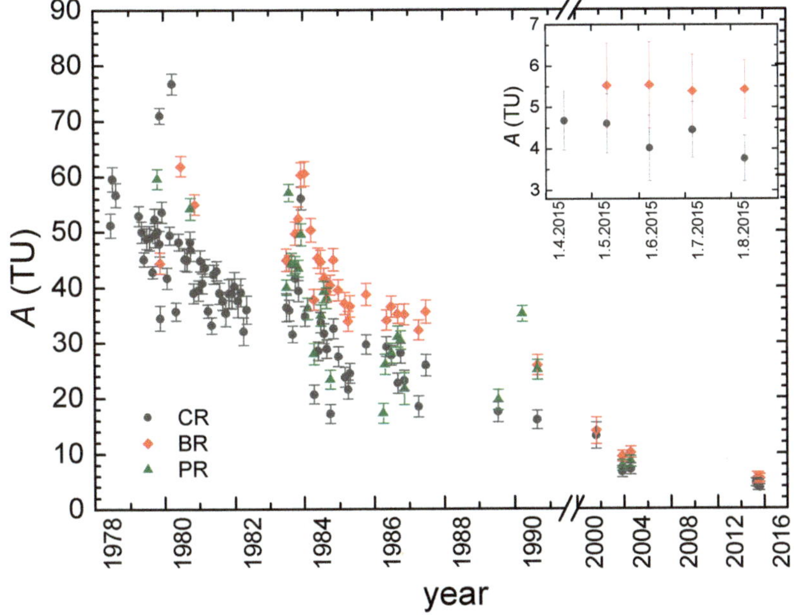

Figure 8. Tritium activity concentration in groundwater (CR, BR, and PR springs), 1978–2016. Data for the CR and BR springs in the spring–summer of 2015 are shown in the insert.

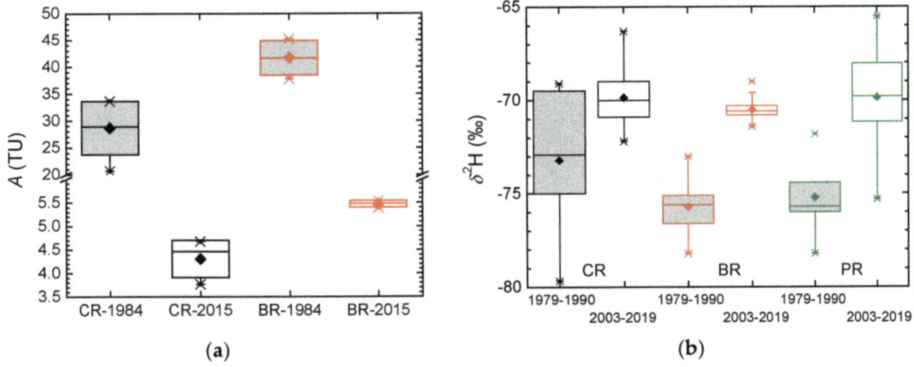

Figure 9. (**a**) Comparison of the tritium activity concentration in the Crna Rijeka (CR) and Bijela Rijeka (BR) springs in 1984 and 2015; (**b**) comparison of the δ^2H values in the CR, BR, and Plitvica River (PR) springs in the two periods, 1979–1990 and 2003–2019.

The range in the $\delta^{18}O$ in precipitation was 14.2‰ and in δ^2H it was 110.1‰ (Table S2). However, the ranges in the δ values in groundwater (at springs) were much lower, at less than 1‰ in $\delta^{18}O$ and less than 10‰ in δ^2H (Figure 10). Again, we compared the δ^2H values in each spring in the two different periods, 1979–1990 and 2003–2019 (Figure 9b). It can be noticed that, in each period, all three springs have similar values, and Bijela Rijeka has the narrowest range in both periods. Interestingly, all three springs have higher δ^2H values in the second period, 2003–2019, by 3–5‰. Earlier, we noticed an increase in temperature in the area, and this increase in the δ^2H values in spring can be attributed to the increase in temperature of 1.2 to 1.9 °C when the two periods are concerned, which is in agreement with the increase in temperature of 0.06 °C per year (Figure 3b).

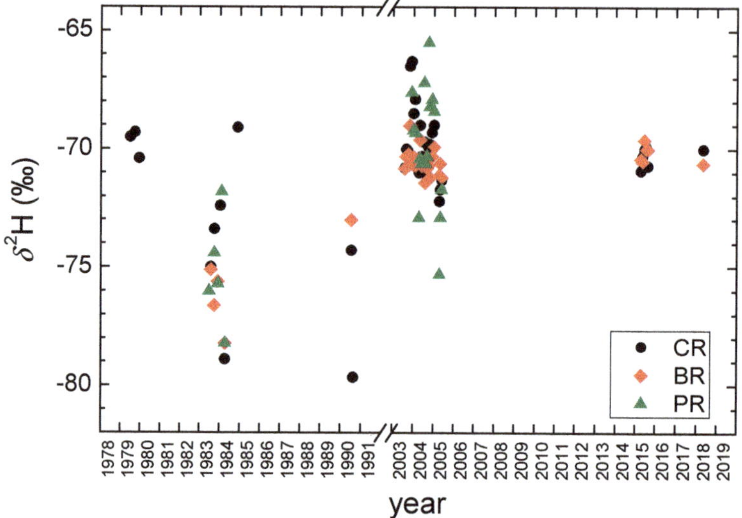

Figure 10. The δ^2H values in groundwater (CR, BR, and PR springs), 1978–2019. Note the break between 1991 and 2003.

Although the isotope data presented above and the physico-chemical and carbon isotope data from previous studies [7,33,46] indicate that the three springs originated from different hydrogeological units, and in principle different groundwater water lines should be obtained, the δ^2H and δ^{18}O data in the three springs cluster together along the LMWL line for the Plitvice Lakes, and the Groundwater Water Line (GWL) is determined for all the springs (Figure 11, Table 4):

$$\delta^2\text{H} = (5.7 \pm 0.7)\,\delta^{18}\text{O} + (-10 \pm 7),\ n = 31,\ r = 0.84. \tag{4}$$

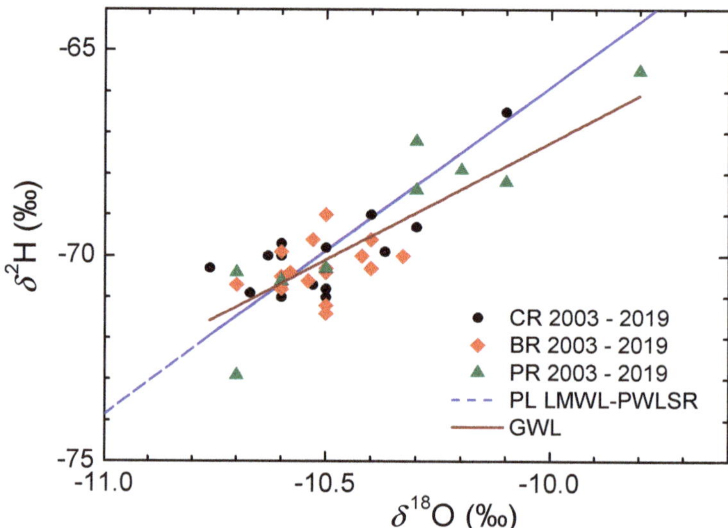

Figure 11. Groundwater water line (GWL) for the three springs, all data after 2000, GWL is given by Equation (4). Comparison with the LMWL$_{\text{PWLSR}}$ for the Plitvice Lakes (Table 3).

Table 4. Values of the slopes ($a \pm \sigma_a$) and intercepts ($b \pm \sigma_b$) for the groundwater line (GWL), surface water lines (SWL), and lake water lines (LWL).

Type of Line	Slope	Intercept	n	r	Ref.
GWL, 1983–1984	7.8	8.5	13		[7]
RGWL_karst	7.0	4.2	340	$R^2 = 0.98$	[30]
GWL, 2000–2018	5.7 ± 0.7	-10.1 ± 7.3	31	0.84	
SWL, 2003–2005	5.0 ± 0.6	-17.3 ± 5.9	35	0.86	[10]
SWL, 2011–2014	6.3 ± 0.3	-3.4 ± 2.7	87	0.94	
LWL, 2011–2014	5.8 ± 0.3	-9.4 ± 3.0	42	0.95	

The GWL (Equation (4)) has both a lower slope and a lower intercept than the LMWL (Table 3), which may be a consequence of water evaporation during its flow in karst aquifers, but also due to a very narrow range of the δ^2H values for BR. The regional groundwater line (RGWL) for the Croatian karst area, based on 340 karst springs, revealed a slope of 7.0 and an intercept of 4.2 (Table 4) [30].

The physico-chemical parameters in the spring waters (temperature, pH, ion concentrations) were also very constant throughout year, and all these together indicated a good mixing of water, uniformly recharged during the year in ground water karst aquifers, and the mean residence time of several years [7,33,46]. The conclusion was confirmed by applying the tritium activity concentrations in groundwater/springs for the determination of the Mean Residence Time (MRT) by the exponential model, which supposed a complete mixing of a new recharge with the already existing water in the aquifer [6–8,70]. The measured A values for the Zagreb precipitation were used as the input data. The MRT values of different karst springs were very short [8], ranging between 1 and 4 years on average. Among the three springs in the Plitvice Lakes area, the Crna Rijeka spring (CR) had the shortest MRT (2 years) and the Bijela Rijeka spring (BR) had the longest of about 4 years. All these data corroborate the previously obtained isotope data: the longer the MRT, the smaller the range of the isotope composition of groundwater (Figure 9).

The extreme climatological characteristics in 1983 and 1984 that enabled the determination of the relative contributions of base-flow and precipitation in the three karst springs are of special interest here. Summer and fall/autumn 1983 were extremely warm and dry, so the aquifers were not completely recharged, and as a result the older water with higher tritium activity concentrations appeared at the springs (Figure 12a). In the following autumn, abundant precipitation was recorded, as well as a high amount of snow in the winter months (February and March 1984), and these caused lower tritium activities in springs in the spring of 1984 (Figure 12a). There are no available P and T data at the Plitvice Lakes in 1983 and 1984. However, the data for Zagreb could help describe the climatological conditions in these years. The mean annual temperature in Zagreb in 1983 was 12.1 °C, which is higher than that in 1982 (11.7 °C) and in 1984 (10.9 °C), and also higher than the average temperature for the 1980–1985 period of 11.2 ± 0.9 °C [23]. The precipitation amount in 1983 was 755 mm, which is lower than that in 1982 (805 mm) and in 1984 (897 mm), and also lower than the mean P in the 1980–1985 period (843 ± 67 mm) [23]. The flow rates at all three springs in autumn 1983 were very low (the lowest flow rates in November 1983: 0.36, 0.018, and 0.12 m^3 s^{-1} for CR, BR, and PR, respectively), much lower than the average flow rates in 1983 (1.327, 0.387, and 0.407 m^3 s^{-1} for CR, BR, and PR, respectively) or the long-term averages (1.41, 0.466, and 0.668 m^3 s^{-1}). At all three springs, the highest flow rates were measured in April 1984 (9.12, 1.81, and 13.3 m^3 s^{-1} for CR, BR, and PR, respectively) [45] (Figure 12b).

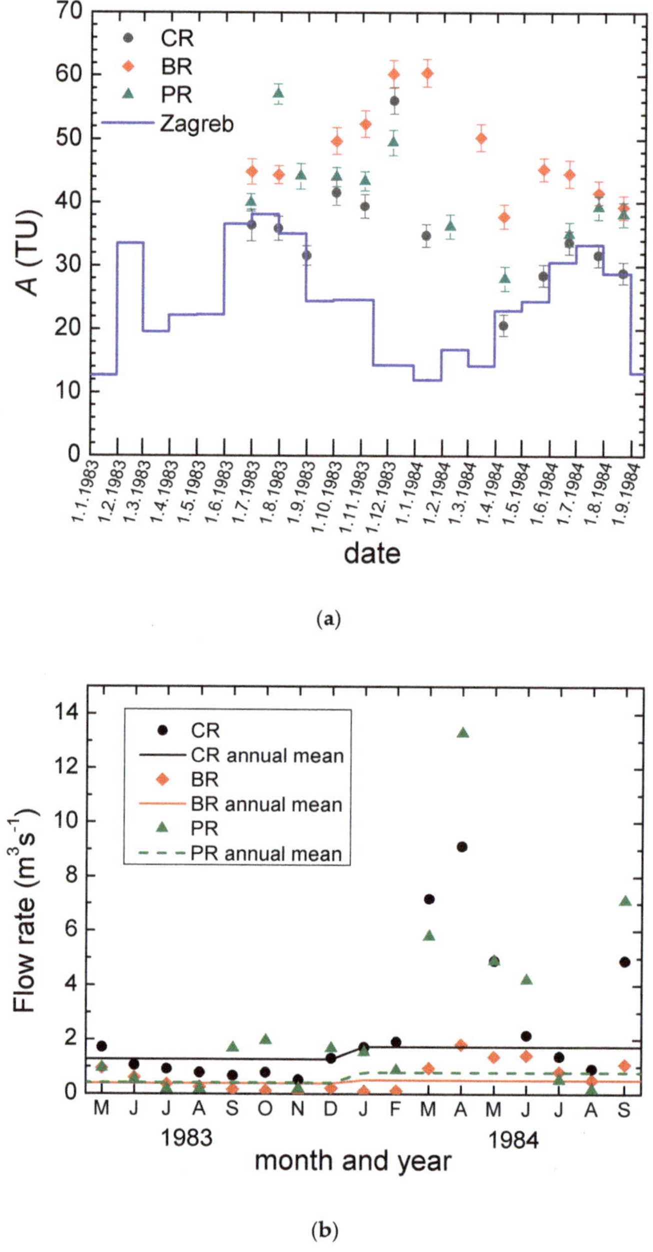

Figure 12. (**a**) Tritium activity concentration in the CR, BR, and PR springs in 1983 and 1984, when extreme hydro-meteorological conditions appeared; (**b**) flow rates at the CR, BR, and PR springs during 1983–1984. Symbols refer to the maximal flow rates in respective months, and lines represent the mean annual flow rates in 1983 and 1984.

The exponential model was applied for MRT calculation under the conditions of the highest and lowest A values in springs, although strictly the exponential model is not valid for non-uniform recharge. The calculated MRT from the higher A values in summer was doubled, for CR to 4 years and for BR to 8 years, in comparison with the previously determined 2 and 4 years, respectively. A significant decline in the tritium activity concentrations in springs was observed after abundant autumn and winter precipitation with lower tritium activity concentrations [7,8,35]. The lowest A values were observed at all three springs in April 1984 after intensive snow melting, and the corresponding MRTs were lower, at ~1 year and ~2 years for CR and BR, respectively. The lowest δ^2H values in spring in April 1984 (Figure 10) reflected the high contribution of winter precipitation. Knowing the A of precipitation, and the A in springs at the highest values (during the dry period) and at the lowest ones (April 1984), a fraction of precipitation input in the groundwater was estimated. The fraction of new (precipitation) water in April 1984 was the highest in CR, at about 90%, and was about 60% in PR, while BR consisted of approximately equal proportions of old groundwater and new input (50% each).

A similar conclusion about the mixing of quick and slow components in the springs of the Plitvice Lakes area was obtained by multiple tracers [10,11]. To obtain the MRTs, a multi-tracer lumped parameter modelling approach was applied using the time series of stable isotopes (^{2}H and ^{18}O) and tritium with a monthly resolution, as well as noble gases and CFCs. The average MRT of most springs was less than 5 years [10,11]. Two components of the groundwater flow were distinguished: the quick flow in the conduit network with an MRT of up to 0.5 years, and the slow component of the matrix of a fissured-porous aquifer with an MRT of up to 28 years [10,11].

3.4. Surface and Lake Waters

A relatively large set of δ^2H data in surface waters in the period 2003–2005 can be found in [10], but the number of δ^{18}O data is significantly lower (Table S1). The data show seasonal fluctuations, with maxima in late summer and minima in late winter (Figure 13a). However, the amplitude in δ^2H of less than 10‰ is much smaller than the amplitude of the δ^2H values in precipitation (about 110‰, Figure 6), and reflects the isotope composition of groundwaters (Figure 6). The maxima are not observed at all locations at the same time due to the retention of water in the two large lakes (Lake Prošćansko and Lake Kozjak). The two subsequent locations, LB before Lake Kozjak and KzB at the end of Lake Kozjak, show a very similar seasonal pattern, with a delay of approximately two months. The data are almost identical at locations KzB and KoB, although four smaller lakes are situated in between, indicating a fast flow of water through the small lakes. The average values increase in the downstream direction (Figure 13b) due to the evaporation of the surface waters, resulting in a Surface Water Line (SWL) with a lower slope (5.0 ± 0.6) and intercept (−17 ± 6) than the PL LMWL$_{PWLSR}$ line (Table 4).

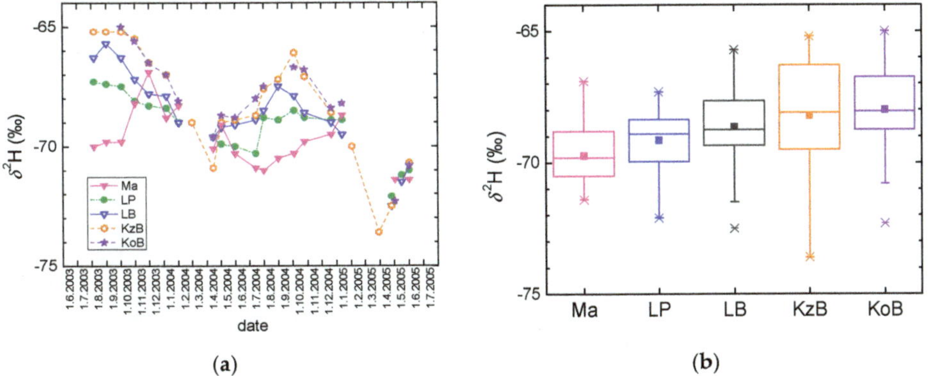

(a)　　　　　　　　　　　　(b)

Figure 13. (a) δ^2H values in the surface waters, 2003–2005, data from [10]; (b) box-plots of δ^2H values at the same locations.

The more recent sampling campaign (between 2011 and 2014) included also lake water from sediment traps at certain water depths (Table 1, Figure 14a, Figure S1). No difference between lake water at selected depths and surface waters at the closest sampling sites is observed when the mean values are concerned, as can be seen from basic statistics data presented in Table S3. A slight increase in both the mean δ^2H and δ^{18}O values and their seasonal variations is observed for locations along the water course. The δ^{18}O increased from −10.7 ± 0.1‰ at Matica to −10.3 ± 0.2‰ at the Korana River–Sastavci (KoS) (Figure S2). Similarly, the δ^2H increased from −71.4 ± 1.1‰ at Matica to −69.0 ± 1.7‰ at KoS (Figure 14b). Both the δ^2H and δ^{18}O are slightly lower at the final location in this study, KoB, probably due to some local groundwater input along the Korana River course. There is no significant difference between the isotope composition of the lake water and the corresponding surface water.

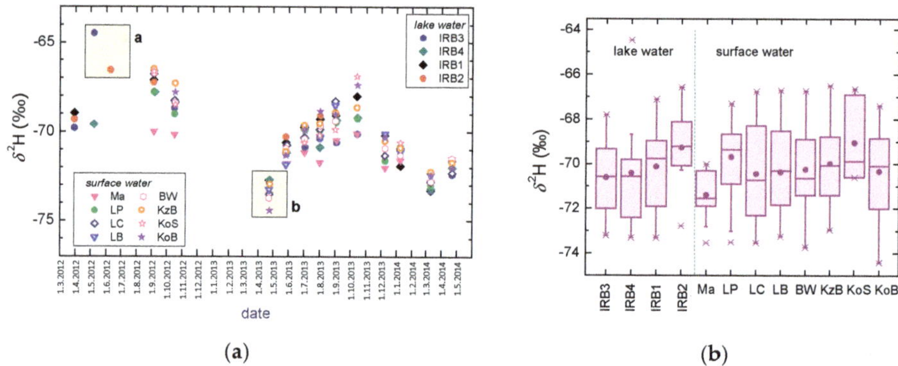

Figure 14. (**a**) Seasonal variation in the δ^2H of surface and lake waters for 2012–2014. For an explanation of groups (**a**,**b**), see text. (**b**) Box-plots of δ^2H values at the same locations.

All the surface waters (2011–2014) lie on the line δ^2H = (6.40 ± 0.26) δ^{18}O–(3.4 ± 2.7), r = 0.94 (Surface Water Line (SWL) 2011–2014, Table 4, Figure 15), and the lake water from traps on the line δ^2H = (5.8 ± 0.3) δ^{18}O–(9.4 ± 3.0), r = 0.95 (Lake Water Line (LWL), Table 4, Figure 15). The SWL line for 2011–2014 is slightly steeper that the SWL for 2003–2005 (Table 4), which could be explained by the different locations and differences in climatological conditions (2003 was the warmest year in Zagreb and at the Plitvice Lakes in those periods [45]). Additionally, there is no significant difference observed between the SWL and LWL (2011–2014), although both have lower slopes than the LMWL$_{\text{PWLSR}}$ obtained for the PL area (Figure 15). The lower slopes and intercepts of both SWL and LWL relations compared to the LMWL indicate the influence of the evaporation of surface waters, which is more pronounced in big lakes.

To justify this conclusion, for all the surface and lake waters from the 2011–2014 sampling campaign we prepared the relations δ^2H vs. δ^{18}O, and the slopes and intercepts are shown in Table S4. A general decrease in both the slopes and intercepts in the downstream direction is observed. The lowest values are found in the biggest Lake Kozjak (IRB1, IRB2, and KzB).

The tritium activity concentration was determined in the surface water at the locations Matica and Lake Kozjak–Bridges in 2015, at the same time as in the groundwaters at the CR and BR springs (Figure 8, insert). The lowest mean A value was observed at CR (4.3 ± 0.4 TU, n = 5), and the highest at BR (5.5 ± 0.1 TU, n = 4), and the A values the surface waters were in between them (Ma: 4.8 ± 0.2 TU, n = 5, KzB: 5.1 ± 0.2, n = 4) (Figure S3), as one would expect after the merging of the Bijela Rijeka River and the Crna Rijeka River into the Matica River.

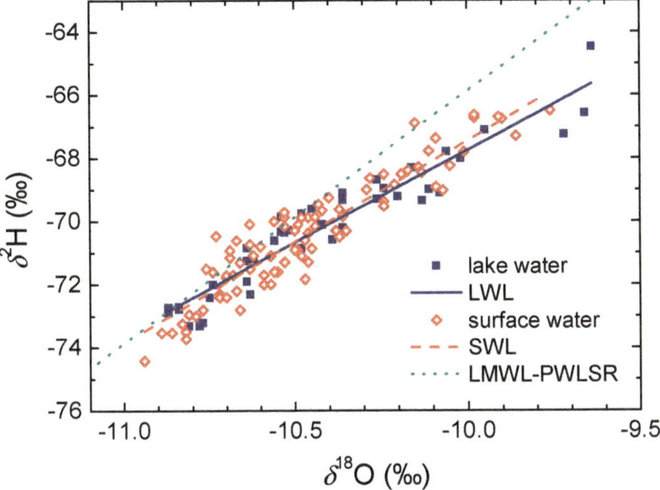

Figure 15. A relation of δ^2H vs. δ^{18}O for the surface and lake waters at the Plitvice Lakes, 2011–2014. SWL and LWL are shown (Table 4), as well as LMWL$_{PWLSR}$ for the Plitvice Lakes.

Although the range of δ^{18}O values in the lake water is not large, systematic data can help in identifying extreme hydrological events, similar to the event in 1983–1984 explained earlier. The relation between δ^{18}O and water temperature (Figure 16) shows a generally increasing trend, as would be expected from the general relation of δ^{18}O in precipitation with temperature [23]. The influence of heavy summer rains and snow melting was observed by a slight increase and decrease, respectively, in the δ^{18}O values compared to the average ("normal") values in both the surface and lake waters. The higher δ^{18}O data points (group **a**, Figure 16) are the consequences of heavy summer rains. For example, in May and June 2012 the monthly precipitation amounts of 90 mm and 144 mm were recorded in Zagreb, with δ^{18}O values of −4.63‰ and −5.32‰, respectively [23]. Unfortunately, data neither for the precipitation amount nor the δ^{18}O in precipitation are available for these months at the PL area. Nevertheless, the effect of the isotopically heavier precipitation is visible in group **a** (Figure 16), as well as in Figure S2. The isotope data of group **b** (Figure 16) are a consequence of the abundant isotopically lighter precipitation (rain and snow) during January–March 2013; the amount of precipitation was 148, 105, and 126 mm in January, February, and March 2013, respectively, with the δ^{18}O values of −12.04‰, −13.73‰, and −11.22‰, respectively (all the data are for Zagreb [23]). The effect of the relatively lower δ^{18}O and δ^2H values was observed in all the surface and lake waters in April 2013 (Figure 14a and Figure S2).

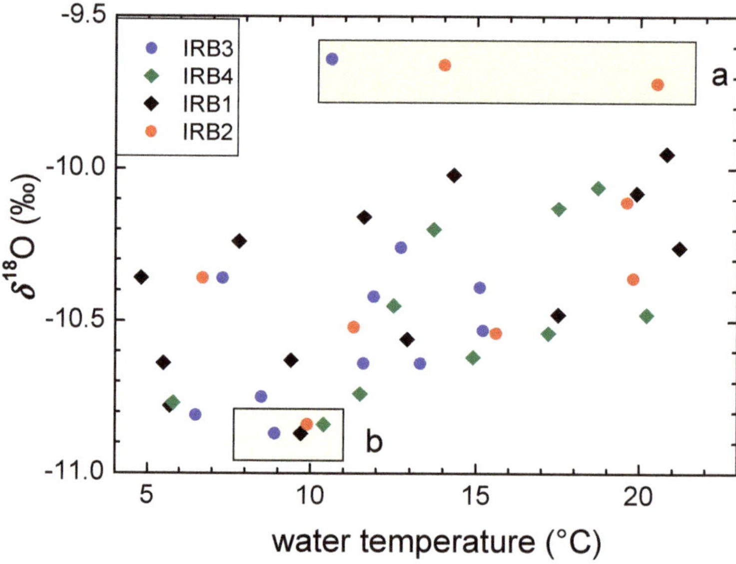

Figure 16. A relation of the $\delta^{18}O$ vs. water temperature for lake waters, 2012–2014. Group **a**: after heavy summer rains; group **b**: after abundant winter precipitation and snow melting.

4. Conclusions

In this paper, we presented an overview of the application of tritium and stable isotopes of water (^2H, ^{18}O) on the characterization of different water bodies (precipitation, groundwater, surface water, lake water) of the karst area of the Plitvice Lakes, Croatia. Various studies were performed over a relatively long time period (1979–2018, with a gap between 1990 and 2001). The available climatological data (amount of precipitation, air temperature) were also analyzed in a search for the evidence of climatological changes. The isotope data were compared to the continuous long-term data record for Zagreb precipitation.

The main conclusion/results of this overview are the following:

- An increase in the mean annual air temperatures of 0.06 °C per year is observed for the period 1986–2019. An increase in the annual precipitation amount is also observed, at about 10 mm per year, and the range of the monthly precipitation amounts is higher in recent years.

- Tritium activity concentration in the Plitvice Lakes precipitation shows characteristics typical for the northern hemisphere. A good correlation of the tritium activity concentration in the precipitation at the Plitvice Lakes with that in the Zagreb precipitation is observed, implying that the tritium data for Zagreb can be safely used for the study of the PL area if/when data do not exist for the PL precipitation.

- The range of the $\delta^{18}O$ in precipitation was about 14‰, and in δ^2H it was about 110‰. Various regression approaches for the determination of LMWL were applied, and the best results were obtained by the RMA and PWLSR approaches, which gave also the best results for the stations Zagreb and Zadar. The LMWL$_{PWLSR}$ for PL is $\delta^2H = (7.97 \pm 0.12)\,\delta^{18}O + (13.8 \pm 1.3)$, $n = 36$, for the period 2003–2006. This LMWL lies in between the LMWL for Zagreb and Zadar.

- The deuterium excess for precipitation at the Plitvice Lakes is higher than that for the Zagreb precipitation. This is caused by the combination of higher altitude and the more intense influence of the Mediterranean air masses. The seasonal pattern in d-excess is similar to that in Zagreb, with higher values in the autumn precipitation due to the higher influence of precipitation of Mediterranean origin.

- The ranges in the δ values in groundwaters (at springs) were much narrower than those in precipitation, at less than 1‰ in $\delta^{18}O$ and less than 8‰ in δ^2H, indicating the good mixing of waters in karst aquifers.

- The higher mean δ values in all three karst springs were observed in recent decades (2003–2019) than in the older period 1979–1990. It can be attributed to the increase in the mean air temperature.

- Different values of mean residence time (MRT) were obtained for three main springs (between 2 years for Crna Rijeka spring and 4 years for Bijela Rijeka spring). Extreme climatological and hydrological conditions in 1983–1984 enabled the estimation of the proportions of precipitation in the spring water: the shorter the MRT, the higher proportions of precipitation.

- The amplitude in the δ^2H of surface and lake water is less than 10‰, and $< 1.4‰$ in $\delta^{18}O$, similar to that in the groundwaters. A slight increase in both the mean δ^2H and $\delta^{18}O$ values and their seasonal variations is observed for locations along the water course due to the evaporation of surface waters. No difference between lake and surface waters is observed if the mean values are concerned. There is no significant difference observed between the Surface Water Line (SWL) and Lake Water Line (LWL) (2011–2014), both having lower slopes than the LMWL$_{PWLSR}$ obtained for the PL area. The stable isotope composition of the surface and lake waters reacts to the extreme hydrological conditions.

The long-term and comprehensive isotope study of different water bodies in the area of the Plitvice Lakes can be an example of how the application of water isotopes (^2H, ^{18}O, ^3H) can help in the characterization of karst aquifers on the regional and global scales. The presented data have also shown what could be the topic of future research. A systematic long-term monitoring of water isotopes should be established, which could, together with the new data from the hydrogeological research, result in more detailed definition on the groundwater flow through the research area and possibly differentiate the aquifers from which the Plitvice Lakes receive the water. The usefulness of the tritium activity concentration in hydrogeological research has become recently of less importance due to the relatively constant mean value of A in precipitation during the last three decades. However, the monitoring of A in precipitation at the Plitvice Lakes may help in distinguishing global from local influences on tritium in precipitation. The further monitoring of the stable isotope data in groundwaters and surface water in relation to the global and local climate changes can give valuable data for the protection of waters, as well as tufa and biota in the area. The development of the hydrogeological conceptual model is also foreseen.

Supplementary Materials: The following are available online at http://www.mdpi.com/2073-4441/12/9/2414/s1, all in a single file: Figure S1: Seasonal variations in $\delta^{18}O$ of surface and lake waters, 2012–2014 period; Figure S2: Box-plot of $\delta^{18}O$ values in lake and surface waters, sorted in downstream direction, 2011–2014 period; Figure S3. Tritium activity concentration in CR and BR springs and in surface water at locations Ma and KzB in 2015; Table S1: Number of isotope data for precipitation, groundwater, surface and lake waters from the Plitvice Lakes area; Table S2: Monthly precipitation amount P, mean monthly temperature T and isotope composition of precipitation ($\delta^{18}O$, δ^2H, deuterium excess, tritium activity concentration A) at the Plitvice Lakes; Table S3: Basic statistics (mean values and standard deviations, minimum, median and maximum values) for $\delta^{18}O$ (shadowed rows) and δ^2H in lake waters (IRB1, IRB2, IRB3, IRB4) and surface waters, 2011–2014 period; Table S4: Slopes and intercepts of relations δ2H vs. δ18O at individual sampling locations of lake waters from traps (IRB1, IRB2, IRB3, IRB4) and surface waters (Ma, LP, LC, LB, BW, KzB, KoS, KoB), 2011–2014 period.

Author Contributions: Conceptualization, data validation and interpretation, writing the manuscript—original draft preparation, final manuscript, visualization—most graphs, and communicating with the journal, I.K.B.; sample collection, tritium activity sample preparation, introduction and participation in discussion, J.B. and A.S.; database management, and LMWL Freeware software application, D.B.; geological setting and visualization, I.L.M.; sample collection, tritium activity sample preparation and interpretation, funding acquisition, N.H.; stable isotope measurement, Z.K. All the co-authors participated in discussions and reviewed the manuscript. All authors have read and agreed to the published version of the manuscript.

Funding: This research received no external funding. In the past, the isotope research on the Plitvice Lakes was part of the following funders and projects: the Ministry of Science and Education (MSE) of the Republic of Croatia, project nos. 00980207 (Natural radioisotopes and processes in gases, 1996–2002), 0098014 (Natural isotopes of low-level activities and the development of instrumentation, 2002–2006), 098-0982709-2741 (Natural radioisotopes in the investigation of Karst ecosystems and dating, 2007–2011); EU project FP5 ICA2-CT-2002-10009 ANTHROPOL.PROT (A study of anthropogenic influence after the war and establishing protection measures in the National Park Plitvice and the Bihać Region at the border area between Croatia and Bosnia and Herzegovina, 2003–2005); and a project financed by the National Park Plitvice Lakes ("An investigation of the influence of forest ecosystems of the National Park Plitvice Lakes on the quality of water and lakes", 2003–2006). The study was supported by the project "The influence of climatic changes and environmental conditions on biologically induced precipitation of tufa and sedimentation process in the Plitvice Lakes", contracted between the Ruđer Bošković Institute and the National Park Plitvice Lakes, 2011–2014.

Acknowledgments: The authors are thankful to former Ruđer Bošković Institute laboratory staff members Bogomil Obelić, Dušan Srdoč, Elvira Hernaus, and Božica Mustač and the present technician Anita Rajtarić, who all took part in sample and data collection. We are thankful to the staff of the National Park Plitvice Lakes for their continuous support and help. We are also thankful to colleagues from the Jožef Stefan Institute in Ljubljana (Jože Pezdič, Polona Vreča, Sonja Lojen, Nives Ogrinc, Silva Perko, Zdenka Trkov, and Stojan Žigon) for stable isotope measurements from the period 1980–2003. Meteorological data were obtained on request (free of charge) from the Croatian Meteorological and Hydrological Service. We thank Catherine E. Hughes and Jagoda Crawford for help in getting LMWL Freeware software and for discussions of the formulae used.

Conflicts of Interest: The authors declare no conflict of interest. The funders had no role in the design of the study; in the collection, analysis, or interpretation of data; in the writing of the manuscript; or in the decision to publish the results.

Abbreviations

CMHS	Croatian Meteorological and Hydrological Service
GMWL	Global Meteoric Water Line
GNIP	Global Network of Isotopes in Precipitation
GPC	Gas Proportional Counting
GWL	Groundwater Water Line
IAEA	International Atomic Energy Agency
IRMS	Isotope Ratio Mass Spectrometry
LMWL	Local Meteoric Water Line
LSC-EE	Liquid Scintillation Counting with Electrolytic Enrichment
LWL	Lake Water Line
MA	Major Axis least squared regression
MRT	mean residence time
OLSR	Ordinary Least Square Regression
PL	Plitvice Lakes
PWLSR	Precipitation weighted OLSR
PWMA	Precipitation weighted MA
PWRMA	Precipitation weighted RMA
RBI	Ruđer Bošković Institute
RMA	reduced major axis regression
rmSSEav	average of the root mean square sum of squared errors
SMOW	Standard Mean Ocean Water
SWL	Surface Water Line
TU	Tritium Unit
UNESCO	The United Nations Educational, Scientific and Cultural Organisation
WMO	World Meteorological Organization

References

1. Bradley, C.; Baker, A.; Jex, C.N.; Leng, M.J. Hydrological uncertainties in the modelling of cave drip-water $\delta^{18}O$ and the implications for stalagmite palaeoclimate reconstructions. *Quat. Sci. Rev.* **2010**, *29*, 2201–2214. [CrossRef]

2. Biondić, B.; Biondić, R.; Dukanić, F. Protection of karst aquifers in the Dinarides in Croatia. *Environ. Geol.* **1998**, *34*, 309–319. [CrossRef]

3. Fajković, H.; Hasan, O.; Miko, S.; Juračić, M.; Mesić, S.; Prohić, E. Vulnerability of the karst area related to potentially toxic elements. *Geol. Croat.* **2011**, *64*, 41–48. [CrossRef]

4.   Terzić, J.; Marković, T.; Lukač Reberski, J. Hydrogeological properties of a complex Dinaric karst catchment: Miljacka Spring case study. *Environ. Earth Sci.* **2014**, *72*, 1129–1142. [CrossRef]

5.   Kazakis, N.; Chalikakis, K.; Mazzilli, N.; Ollivier, C.; Manakos, A.; Voudouris, K. Management and research strategies of karst aquifers in Greece: Literature overview and exemplification based on hydrodynamic modelling and vulnerability assessment of a strategic karst aquifer. *Sci. Total Environ.* **2018**, *643*, 592–609. [CrossRef]

6.   Małoszewski, P.; Rauert, W.; Stichler, W.; Herrmann, A. Application of flow models in an alpine catchment area using tritium and deuterium data. *J. Hydrol.* **1983**, *66*, 319–330. [CrossRef]

7.   Srdoč, D.; Horvatinčić, N.; Obelić, B.; Krajcar Bronić, I.; Sliepčević, A. Procesi taloženja kalcita u krškim vodama s posebnim osvrtom na Plitvička jezera (Calcite deposition processes in karstwaters with special emphasis on the Plitvice Lakes, Yugoslavia). *Carsus Iugoslaviae* **1985**, *11*, 101–204.

8.   Srdoč, D.; Krajcar Bronić, I. The application of stable and radioactive isotopes in karst water research. *Naš Krš* **1986**, *12*, 37–47.

9.   Małoszewski, P.; Zuber, A. Lumped parameter models for the interpretation of environmental tracer data. In *Manual on Mathematical Models in Isotope Hydrology*; IAEA-TECDOC-910; IAEA: Vienna, Austria, 1996; pp. 9–58.

10.   Babinka, S. Multi-tracer study of karst waters and lake sediments in Croatia and Bosnia-Herzegovina: Plitvice Lakes National Park and Bihać area. Ph.D. Thesis, Rheinischen Friedrich-Wilhelms-Universität, Bonn, Germany, 2007; p. 167.

11.   Babinka, S.; Obelić, B.; Krajcar Bronić, I.; Horvatinčić, N.; Barešić, J.; Kapelj, S.; Suckow, A. Mean residence time of water from springs of the Plitvice Lakes and Una River area. In *Advances in Isotope Hydrology and Its Role in Sustainable Water Resources Management (HIS-2007), Proceedings of a Symposium, Vienna, Austria, 21–25 May 2007*; IAEA-CN-15/45; IAEA: Vienna, Austria, 2007; pp. 327–336. Available online: https://www-pub.iaea.org/MTCD/publications/PDF/Pub1310_web.pdf (accessed on 5 August 2020).

12.   Horvatinčić, N.; Kapelj, S.; Sironić, A.; Krajcar Bronić, I.; Kapelj, J.; Marković, T. Investigation of water resources and water protection in the karst area of Croatia using isotopic and geochemical analyses. In *Advances in Isotope Hydrology and Its Role in Sustainable Water Resources Management (HIS-2007), Proceedings of a Symposium, Vienna, Austria, 21–25 May 2007*; IAEA: Vienna, Austria, 2007; pp. 295–304. ISBN 978-92-0-110207-2. Available online: https://www-pub.iaea.org/MTCD/publications/PDF/Pub1310Vol2_web.pdf (accessed on 5 August 2020).

13.   Häusler, H.; Frančišković-Bilinski, S.; Rank, D.; Stadler, P.; Bilinski, H. Preliminary results from stable isotope investigations in the Kupa Drainage Basin (Western Croatia). In *BALWOIS 2012 Conference Web-Site, Ohrid, Macedonia, 28 May–3 June 2012*; Morell, M., Popovska, C., Morell, O., Stojov, V., Eds.; BALWOIS: Ohrid, Macedonia, 2012.

14.   Mandić, M.; Bojić, D.; Roller-Lutz, Z.; Lutz, H.O.; Krajcar Bronić, I. Note on the spring region of Gacka River (Croatia). *Isot. Environ. Health Stud.* **2008**, *44*, 201–208. [CrossRef]

15.   Cartwright, I.; Morgenstern, U. Transit times from rainfall to baseflow in headwater catchments estimated using tritium: The Ovens River, Australia. *Hydrol. Earth Syst. Sci.* **2015**, *19*, 1–15. [CrossRef]

16.   Nikolov, J.; Krajcar Bronić, I.; Todorović, N.; Barešić, J.; Petrović Pantić, T.; Marković, T.; Bikit-Schroeder, K.; Stojković, I.; Tomić, M. A survey of isotopic composition (^{2}H, ^{3}H, ^{18}O) of groundwater from Vojvodina. *J. Radioanal. Nucl. Chem.* **2019**, *320*, 385–394. [CrossRef]

17.   Parlov, J.; Kovač, Z.; Nakić, Z.; Barešić, J. Using Water Stable Isotopes for Identifying Groundwater Recharge Sources of the Unconfined Alluvial Zagreb Aquifer (Croatia). *Water* **2019**, *11*, 2177. [CrossRef]

18.   Madonia, P.; Cangemi, M.; Oliveri, Y.; Germani, C. Hydrogeochemical Characters of Karst Aquifers in Central Italy and Relationship with Neotectonics. *Water* **2020**, *12*, 1926. [CrossRef]

19.   Vreča, P.; Krajcar Bronić, I.; Horvatinčić, N.; Barešić, J. Isotopic characteristics of precipitation in Slovenia and Croatia: Comparisom of continental and maritime stations. *J. Hydrol.* **2006**, *330*, 457–469. [CrossRef]

20.   IAEA/WMO. Global Network of Isotopes in Precipitation. The GNIP Database. Available online: https://nucleus.iaea.org/wiser (accessed on 22 November 2019).

21.   Rozanski, K.; Araguás-Araguás, L.; Gonfiantini, R. Isotopic patterns in modern global precipitation. *Geophys. Monogr. Ser.* **1993**, *78*, 1–36. [CrossRef]

22.   Rozanski, K.; Gonfiantini, R.; Araguas-Araguas, L. Tritium in the global atmosphere: Distribution patterns and recent trends. *J. Phys. G Nucl. Part. Phys.* **1991**, *17*, 523–536. [CrossRef]

23. Krajcar Bronić, I.; Barešić, J.; Borković, D.; Sironić, A.; Lovrenčić Mikelić, I.; Vreča, P. Long-Term Isotope Records of Precipitation in Zagreb, Croatia. *Water* **2020**, *12*, 226. [CrossRef]

24. Bognar, A.; Faivre, S.; Buzjak, N.; Pahernik, M.; Bočić, N. Recent Landform Evolution in the Dinaric and Pannonian Regions of Croatia. In *Recent Landform Evolution*; Lóczy, D., Stankoviansky, M., Kotarba, A., Eds.; Springer: Berlin/Heidelberg, Germany, 2012; pp. 313–344. [CrossRef]

25. Surić, M.; Czuppon, G.; Lončarić, R.; Bočić, N.; Lončar, N.; Bajo, P.; Drysdale, R.N. Stable Isotope Hydrology of Cave Groundwater and its Relevance for Speleothem-Based Palaeoenvironmental Reconstruction in Croatia. *Water* **2020**, *12*, 2386. [CrossRef]

26. Bonacci, O. Karst hydrogeology/hydrology of Dinaric chain and isles. *Environ. Earth Sci.* **2015**, *74*, 37–55. [CrossRef]

27. Bonacci, O.; Terzić, J.; Roje-Bonacci, T.; Frangen, T. An Intermittent Karst River: The Case of the Čikola River (Dinaric Karst, Croatia). *Water* **2019**, *11*, 2415. [CrossRef]

28. Garašić, M.; Garašić, D. The new insights about deep karst springs in the Dinaric karst of Croatia. In *Proceedings of the 17th International Congress of Speleology—Volume 2, The 17th International Congress of Speleology, Sydney, NSW, Australia, 23–29 July 2017*; Moore, K., White, S., Eds.; Australian Speleological Federation Inc and Speleo: Sydney, Australia, 2017; pp. 26–29.

29. Bencetić Klaić, Z.; Rubinić, J.; Kapelj, S. Review of research on Plitvice Lakes, Croatia in the fields of meteorology, climatology, hydrology, hydrogeochemistry and physical limnology. *Geofizika* **2018**, *35*, 189–278. [CrossRef]

30. Brkić, Ž.; Kuhta, M.; Hunjak, T.; Larva, O. Regional Isotopic Signatures of Groundwater in Croatia. *Water* **2020**, *12*, 1983. [CrossRef]

31. Horvatinčić, N.; Krajcar Bronić, I.; Obelić, B. Diferences in the ^{14}C age, δ^{13}C and δ^{18}O of Holocene tufa and speleothem in the Dinaric Karst. *Palaeogeogr. Palaeoclimatol. Palaeoecol.* **2003**, *193*, 139–157. [CrossRef]

32. Horvatinčić, N.; Sironić, A.; Barešić, J.; Sondi, I.; Krajcar Bronić, I.; Borković, D. Mineralogical, Organic and Isotopic Composition as Palaeoenvironmental Records in the Lake Sediments of Two Lakes, the Plitvice Lakes, Croatia. *Quat. Int.* **2018**, *494*, 300–313. [CrossRef]

33. Sironić, A.; Barešić, J.; Horvatinčić, N.; Brozinčević, A.; Vurnek, M.; Kapelj, S. Changes in the geochemical parameters of karst lakes over the past three decades—The case of Plitvice Lakes, Croatia. *Appl. Geochem.* **2017**, *78*, 12–22. [CrossRef]

34. Horvatinčić, N. Radiocarbon and tritium measurements in water samples and application of isotopic analyses in hydrology. *Fizika* **1980**, *12*, 201–218.

35. Horvatinčić, N.; Krajcar Bronić, I.; Pezdič, J.; Srdoč, D.; Obelić, B. The distribution or radioactive (^{3}H, ^{14}C) and stable (^{2}H, ^{18}O) isotopes in precipitation, surface and groundwaters during the last decade in Yugoslavia. *Nucl. Instrum. Methods Phys. Res. B* **1986**, *1*, 550–553. [CrossRef]

36. Krajcar Bronić, I.; Horvatinčić, N.; Obelić, B. Two decades of environmental isotope records in Croatia: Reconstruction of the past and prediction of future levels. *Radiocarbon* **1998**, *40*, 399–416. [CrossRef]

37. Krajcar Bronić, I.; Vreča, P.; Horvatinčić, N.; Barešić, J.; Obelić, B. Distribution of hydrogen, oxygen and carbon isotopes in the atmosphere of Croatia and Slovenia. *Arch. Ind. Hyg. Toxicol.* **2006**, *57*, 23–29.

38. Biondić, B.; Biondić, R.; Meaški, H. The conceptual hydrogeological model of the Plitvice Lakes. *Geol. Croat.* **2010**, *63*, 195–206. [CrossRef]

39. Biondić, R.; Meaški, H.; Biondić, B. Hydrogeology of the sinking zone of the Korana River downstream of the Plitvice Lakes, Croatia. *Acta Carsol.* **2016**, *45*, 43–56. [CrossRef]

40. Rubinić, J.; Zwicker, G.; Dragičević, N. Contribution to knowing Plitvice Lakes hydrology—lake water level variation dynamics and significant changes. In Proceedings of the Meeting Hydrological Measurements and Data Processing ("Hidrološka mjerenja i obrada podataka"), NP Plitvice Lakes, Croatia, 26–28 November 2008; pp. 207–230, (In Croatian, English abstract).

41. Polšak, A.; Juriša, M.; Šparica, M.; Šimunić, A. *Basic Geological Map of SFRJ, Sheet Bihać, 1:100 000*; Federal Geological Office: Belgrade, Yugoslavia, 1976.

42. Halamić, J. *Explanatory Notes of the Geological Map of the Republic of Croatia 1:300.000*; Velić, I., Vlahović, I., Eds.; Croatian Geological Institute: Zagreb, Croatia, 2009; p. 141.

43. Köppen, W. Das geographische System der Klimate. In *Handbuch der Klimatologie*; Köppen, W., Geiger, G., Eds.; Gebrüder Borntraeger: Berlin, Germany, 1936; pp. 1–44.

44. Peel, M.C.; Finlyanson, B.L.; McMahon, T.A. Updated world map of the Köppen-Geiger climate classification. *Hydrol. Earth Syst. Sci.* **2007**, *11*, 1633–1644. [CrossRef]

45. CMHS, 2020 Croatian Meteorological and Hydrological Service. Available online: https://meteo.hr/index_en.php (accessed on 25 August 2020).

46. Sironić, A.; Krajcar Bronić, I.; Horvatinčić, N.; Barešić, J.; Borković, D.; Vurnek, M.; Lovrenčić Mikelić, I. Carbon isotopes in dissolved inorganic carbon as tracers of carbon sources in karst waters of the Plitvice Lakes, Croatia. *Geol. Soc. Lond.* **2020**. [CrossRef]

47. Vreča, P.; Krajcar Bronić, I.; Leis, A.; Demšar, M. Isotopic composition of precipitation at the station Ljubljana (Reaktor), Slovenia—period 2007–2010. *Geologija* **2014**, *57*, 217–230. [CrossRef]

48. Morrison, J.; Brockwell, T.; Merren, T.; Fourel, F.; Phillips, A.M. On-line high-precision stable hydrogen isotopic analyses on nanoliter water samples. *Anal. Chem.* **2001**, *73*, 3570–3575. [CrossRef]

49. Coplen, T.B.; Wassenaar, L.I. LIMS for Lasers for achieving long-term accuracy and precision of δ^2H, δ^{17}O, and δ^{18}O of waters using laser absorption spectrometry. *Rapid Commun. Mass Spectrom.* **2015**, *29*, 2122–2130. [CrossRef] [PubMed]

50. Krajcar Bronić, I.; Obelić, B.; Srdoč, D. The simultaneous measurement of tritium activity and background count rate in a proportional counter by the Povinec method: Three years experience at the Rudjer Bošković Institute. *Nucl. Instrum. Methods Phys. Res.* **1986**, *17*, 498–500. [CrossRef]

51. Barešić, J.; Krajcar Bronić, I.; Horvatinčić, N.; Obelić, B.; Sironić, A.; Kožar-Logar, J. Tritium activity measurement of water samples using liquid scintillation counter and electrolytical enrichment. In Proceedings of the 8th Symposium of the Croatian Radiation Protection Association, Krk, Croatia, 13–15 April 2011; Krajcar Bronić, I., Kopjar, N., Milić, M., Branica, G., Eds.; HDZZ: Zagreb, Croatia, 2011; pp. 461–467.

52. Coplen, T.B. Guidelines and recommended terms for expression of stable isotope-ratio and gas-ratio measurement results. *Rapid Commun. Mass Spectrom.* **2011**, *25*, 2538–2560. [CrossRef] [PubMed]

53. IAEA. *Stable Isotope Hydrology: Deuterium and Oxygen-18 in Water Cycle*; Technical Reports Series 210; Gat, J.R., Gonfiantini, R., Eds.; IAEA: Vienna, Austria, 1981; p. 339.

54. Brand, W.A.; Coplen, T.B.; Vogl, J.; Rosner, M.; Prohaska, T. Assessment of International Reference Materials for Isotope-Ratio Analysis (IUPAC Technical Report). *Pure Appl. Chem.* **2014**, *86*, 425–467. [CrossRef]

55. Craig, H. Isotope variations in meteoric waters. *Science* **1961**, *133*, 1702–1703. [CrossRef]

56. Dansgaard, W. Stable isotopes in precipitation. *Tellus* **1964**, *16*, 436–468. [CrossRef]

57. Dansgaard, W. The isotope composition of natural waters. *Medd. Grønland* **1961**, *165*, 120.

58. Mook, W.G. *Environmental Isotopes in the Hydrological Cycle, Principles and Applications, Volumes I, IV and V*; Technical Documents in Hydrology 39; IAEA-UNESCO: Paris, France, 2001.

59. Gat, J.R.; Carmi, I. Evolution of the Isotopic Composition of Atmospheric Waters in the Mediterranean Sea Area. *J. Geophys. Res.* **1970**, *75*, 3039–3048. [CrossRef]

60. Gat, J.R.; Shemesh, A.; Tziperman, E.; Hecht, A.; Georgopoulos, D.; Basturk, O. The stable isotope composition of waters of the eastern Mediterranean Sea. *J. Geophys. Res.* **1996**, *101*, 6441–6451. [CrossRef]

61. Crawford, J.; Hughes, C.E.; Lykoudis, S. Alternative least squares methods for determining the meteoric water line, demonstrated using GNIP data. *J. Hydrol.* **2014**, *519*, 2331–2340. [CrossRef]

62. Hughes, C.E.; Crawford, J. A new precipitation weighted method for determining the meteoric water line for hydrological applications demonstrated using Australian and global GNIP data. *J. Hydrol.* **2012**, *464*, 344–351. [CrossRef]

63. IAEA. *Statistical Treatment of Environmental Isotopes in Precipitation*; Technical Report Series 331; IAEA: Vienna, Austria, 1992; p. 781.

64. ANSTO. Australian Nuclear Science and Technology Organisation. Local Meteoric Water Line Freeware. 2012. Available online: https://openscience.ansto.gov.au/collection/879 (accessed on 28 November 2019).

65. Dolinar, M.; Vertačnik, G. Spremenljivost temperaturnih in padavinskih razmer v Sloveniji. In *Okolje se Spreminja: Podnebna Spremenljivost Slovenije in Njen Vpliv na Vodno Okolje*; Cegnar, T., Ed.; Ministry of Environment: Ljubljana, Slovenia, 2010; pp. 37–42. ISBN 978-961-6024-55-6.

66. NASA. Available online: https://www.nasa.gov/press-release/nasa-noaa-data-show-2016-warmest-year-on-record-globally/ (accessed on 19 August 2020).

67. Global Temperature Vital Signs—Climate Change: Vital Signs of the Planet. Available online: https://climate.nasa.gov/vital-signs/global-temperature/ (accessed on 19 August 2020).

68. Bastin, J.F.; Clark, E.; Elliott, T.; Hart, S.; van den Hoogen, J.; Hordijk, I.; Ma, H.; Majumder, M.; Manoli, G.; Maschler, J.; et al. Understanding climate change from a global analysis of city analogues. *PLoS ONE* **2019**, *14*, e0217592. [CrossRef]
69. Barešić, J.; Horvatinčić, N.; Krajcar Bronić, I.; Obelić, B.; Vreča, P. Stable isotope composition of daily and monthly precipitation in Zagreb. *Isot. Environ. Health Stud.* **2006**, *42*, 239–249. [CrossRef]
70. Geyh, M.A. Basic studies in hydrology and ^{14}C and ^{3}H measurements. In *General Proceedings of the 24th International Geological Congress, The 24th International Geological Congress, Montreal, QC Canada, 21–20 August 1972*; Armstrong, J.E., Ed.; International Geological Congress: Montreal, QC, Canada, 1973; Volume 11, pp. 227–234.

Article

Stable Isotope Hydrology of Cave Groundwater and Its Relevance for Speleothem-Based Paleoenvironmental Reconstruction in Croatia

Maša Surić [1,*], György Czuppon [2,3], Robert Lončarić [1], Neven Bočić [4], Nina Lončar [1], Petra Bajo [5] and Russell N. Drysdale [6,7]

1. Department of Geography, Center for Karst and Coastal Research, University of Zadar, 23000 Zadar, Croatia; rloncar@unizd.hr (R.L.); nloncar@unizd.hr (N.L.)
2. Institute for Geological and Geochemical Research, RCAES, H-1112 Budapest, Hungary; czuppon@geochem.hu
3. Department of Hydrogeology and Engineering Geology, Institute of Environmental Management, University of Miskolc, H-3515 Miskolc, Hungary
4. Department of Geography, Faculty of Science, University of Zagreb, 10000 Zagreb, Croatia; nbocic@geog.pmf.hr
5. Croatian Geological Survey, 10000 Zagreb, Croatia; pbajo@hgi-cgs.hr
6. School of Geography, Faculty of Science, The University of Melbourne, 3010 Melbourne, Australia; rnd@unimelb.edu.au
7. Laboratoire EDYTEM (UMR CNRS 5204), Universite de Savoie-Mont Blanc, 73000 Chambéry, France
* Correspondence: msuric@unizd.hr

Received: 24 July 2020; Accepted: 23 August 2020; Published: 25 August 2020

Abstract: Speleothems deposited from cave drip waters retain, in their calcite lattice, isotopic records of past environmental changes. Among other proxies, $\delta^{18}O$ is recognized as very useful for this purpose, but its accurate interpretation depends on understanding the relationship between precipitation and drip water $\delta^{18}O$, a relationship controlled by climatic settings. We analyzed water isotope data of 17 caves from different latitudes and altitudes in relatively small but diverse Croatian karst regions in order to distinguish the dominant influences. Drip water $\delta^{18}O$ in colder caves generally shows a greater resemblance to the amount-weighted mean of precipitation $\delta^{18}O$ compared to warmer sites, where evaporation plays an important role. However, during glacial periods, today's 'warm' sites were cold, changing the cave characteristics and precipitation $\delta^{18}O$ transmission patterns. Superimposed on these settings, each cave has site-specific features, such as morphology (descending or ascending passages), altitude and infiltration elevation, (micro) location (rain shadow or seaward orientation), aquifer architecture (responsible for the drip water homogenization) and cave atmosphere (governing equilibrium or kinetic fractionation). This necessitates an individual approach and thorough monitoring for best comprehension.

Keywords: stable isotopes; drip water; speleothem; cave; Croatia

---

1. Introduction

Understanding the mechanisms and extent of past environmental changes is a key consideration for predicting global climate changes. Most impacted and potentially endangered is Earth's Critical Zone (ECZ), the thin surface zone that extends from the upper limit of vegetation to the lower limit of groundwater bodies in which coupled chemical, biological, physical, and geological processes operate together to sustain life [1]. This is an extremely heterogeneous zone whose natural functioning is strongly regulated by climate, i.e., changes in temperature and hydrology and their influence on vegetation and soil evolution [2]. One of the natural archives which confidentially record hydroclimate changes in ECZ are speleothems—secondary carbonate deposits that precipitate in a cave environment from groundwaters (cave drip water) supersaturated with respect to calcite following CO_2-driven dissolution of carbonate bedrock [3]. The importance of speleothems for paleoenvironmental reconstructions through the use of stable isotopes was recognized in the 1960's [4], especially the ability of $\delta^{18}O$ to record cave temperature variations [5]. In the following decades, various other proxies have emerged, such as trace elements (Mg, Sr, Y, Pb, Zn, P, U), organic acids, lipid biomarkers, magnetic susceptibility, laminae thickness and growth rate variations, along with development of very precise and accurate dating techniques (U-Th, U-Pb), which have given great momentum to speleothem science. It has been suggested that, after the dominance of ice and deep-sea cores, the future in Quaternary paleoclimate studies might belong to speleothem science [6].

Where equilibrium deposition can be demonstrated, speleothem $\delta^{18}O$ depends solely on cave air temperature and drip water $\delta^{18}O$ [5,7]. Establishing cave hydroclimate monitoring is critical to assess variability in these two parameters. Although it is not always the case [8], caves are usually considered environments with stable microclimate parameters, with relative humidity (RH) close to 100%, and air temperature equal to the surface mean annual air temperature (MAAT), without seasonal variations. When such conditions are met in particular parts of the cave, underground environmental changes respond to long-term climate changes and are recorded in the isotopic composition of slow-growing speleothems. However, successful interpretation of speleothem records depends on comprehension of the spatial and temporal variability of drip water hydrology and composition [9,10] as the water undergoes complex processes on its way from the atmosphere, through the soil and epikarst and to the cave, with final temperature-dependent fractionation and incorporation into the speleothem calcite lattice [11,12].

Croatia's central position between the western and eastern Mediterranean, at the interface of continental European and maritime Mediterranean influences, together with its variety of landscapes, makes this region interesting in terms of paleoclimate and paleoenvironmental studies based on different approaches and archives, such as: loess sequences, [13–16] lake sediments, [17–20], marine and transitional deposits, [21–23], algal rims [24] and speleothems (see an overview in [25] and references therein). Speleothem-based studies require insight into atmospheric isotope patterns from global to local scale, and variations in precipitation isotope composition in Croatia on a local and regional scale have been discussed in Krajcar Bronić et al., [26–28], Horvatinčić et al., [29,30], Barešić et al., [31] and Vreča et al., [32]. Regional isotopic signatures of groundwater from ca. 100 sites in Croatia and their relationship to precipitation are discussed in Brkić et al., [33].

Here we discuss the outcomes of several cave drip water research campaigns. We examine the response of each cave system to surface precipitation and the transmission of the isotopic signals that are ultimately captured in the speleothems as environmental records. The study encompasses 17 sites with elevations from 32 m to 1550 m above sea level (a.s.l.) and four different climate types (Cfa—temperate with hot summer; Cfb—temperate with warm summer; Csa—temperate dry with hot summers; Df—cold and wet continental, according to Köppen-Geiger climate classification [34]), covering most of the Croatian karst regions. The studies can be divided into three groups according to their duration:

- Systematic sampling with at least one-year of continuous monitoring and monthly sampling of rain and drip water was applied within three research projects that included Manita peć (MP), Strašna peć (SP) and Špilja u Zubu Buljme (ZB) caves [35]; Upper Barać (GB), Lower Barać (DB) and Lower Cerovačka (DC) caves [36]; and Nova Grgosova (NG), Lokvarka (LOK) and Modrič (MOD) caves [37,38].
- Semi-continuous sampling was performed in a study of the Velebita Cave System (VEL), and the Sirena (SIR) and Lukina jama (LUK) pits [39].
- Sporadic and one-off sampling occurred in Medvjeđa špilja (MED), Špilja u Vrdolju (VRD), Kraljičina Spilja (KRA), Strašna peć, Mala špilja (MML) and Velika špilja (VML) caves [40–42], and Modrič Cave [43].

We investigated the dominant factors controlling the transmission of the isotopic signal from precipitation to cave drip water in different climate settings, and how this information can be interpreted in speleothem-based climate reconstructions.

2. Study Area

A total of 43.7% of the Croatian territory is covered by karst [44]. The northern and eastern parts of the country are characterized by several isolated karst patches, while practically the whole western and southern parts belong to the Dinaric classical karst, extending parallel to the eastern Adriatic coast (Figure 1). Due to the temperate location (between 42.2 and 46.2° N) with sufficient precipitation year round (700–3000 mm), tectonically fractured carbonate bedrock has undergone extensive karstification, resulting in the development of a large range of surface and underground karst features, including karst hydrogeology. According to a karst landscape definition based on altitude range and climate (aridity index and mean annual days with snow cover), three (out of four) karst landscapes intertwine in this small region of Croatia: humid hills and plains, Mediterranean medium range mountains and high range mountains (see Figure 4 in [45]).

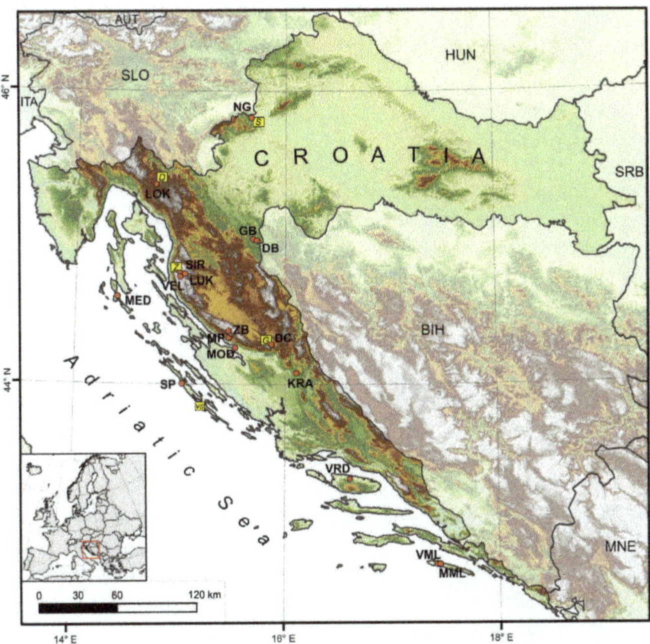

Figure 1. Study area with cave locations: NG—Nova Grgosova, LOK—Lokvarka, GB—Upper Barać, DB—Lower Barać, SIR—Sirena, LUK—Lukina jama, VEL—Velebita, MED—Medvjeđa špilja, ZB—Špilja u Zubu Buljme, MP—Manita peć, DC—Lower Cerovačka, MOD—Modrič, KRA—Kraljičina spilja, SP—Strašna peć, VRD—Špilja u Vrdolju, VML—Velika špilja, MML—Mala špilja. Yellow squares denote meteorological station: S—Samobor 141 m a.s.l.), D—Delnice (681 m a.s.l.), Z—Zavižan (1594 m a.s.l.), G—Gračac (567 m a.s.l.), VS—Vela Sestrica (35 m a.s.l.).

Unofficially, over 10,000 caves are already known in Croatia [46], but this figure is in reality much higher as caves have been discovered in all carbonate geological units, from Upper Palaeozoic to the Neogene. The studied caves (Table 1) can be divided into several groups according to their geographical position and associated climate conditions (Figure 2), which do not necessary imply similar environmental settings of relatively adjacent caves. Most similar are the shallow and low-altitude caves located on the islands and along the coast (MOD, SP, MED, VRD, KRA, MML, VML), then high-mountain caves (ZB, VEL, SIR, LUK) and mid-altitude continental caves which have site-specific climatic settings, regardless of their relative vicinity (NG, LOK, DB, GB, DC). MP cave is, according to its latitude and altitude (570 m a.s.l.), expected to be the most similar to DC cave, but its location on the south-western slopes of the Velebit Mt. makes it more similar to the coastal sites.

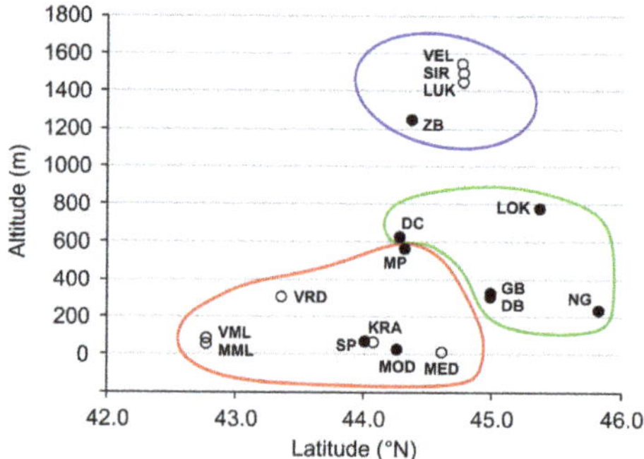

Figure 2. Altitudinal and latitudinal distribution of studied caves clustered according to general climate similarity—coastal and island caves (red), continental mid-altitude caves (green) and high-mountain caves (blue). Closed circles denote caves with systematic monitoring/sampling, while open circles represent caves with sporadic sampling.

Climate and weather patterns over the study area are governed by the seasonal presence of the various pressure systems and air masses. During the summer, the influence of the Azores High prevails in the coastal region and causes dry weather along most of the Croatian coast, while continental Croatia is more susceptible to intrusions of cooler and moist air masses from the Atlantic. During winter, the continental part is often under the influence of Siberian High, which produces cold and dry weather; the coast is under the influence of low pressure systems originating either from the Atlantic Ocean, the Mediterranean or the Adriatic Sea.

Although the study sites are located relatively close to each other, they exhibit quite different climate properties, particularly seasonal precipitation distribution. Most of the locations have precipitation maxima in autumn (October and November) due to the frequent passage of frontal systems within atmospheric lows, with the exception of low-altitude sites in continental Croatia, i.e., NG cave (represented by data from the Samobor meteorological station, Figure 3), which has maximum precipitation in the late summer caused by the combined effect of convective and frontal rain. Sites located in the continental regions, but at higher altitudes (LOK cave represented by the Delnice station), exhibit orographic effects, which cause a surge in the precipitation amount (annual precipitation in Delnice is 2218 mm compared to 1073 mm in Samobor). A similar effect is present in the high-altitude sites (ZB, VEL, SIR, LUK caves, represented by the data from Zavižan station), but with slightly lower precipitation due to the lower air temperatures (MAAT of 5.0 °C Zavižan, and 8.4 °C in Delnice). On the other hand, sites on the leeside of the mountains (DB, GB, DC caves) tend to have somewhat decreased precipitation probably due to their locations in the rain shadow. Coastal and island sites SP, KRA, VRD, VML and MML (represented by Vela Sestrica station) have a typical Mediterranean precipitation distribution with pronounced summer droughts (Csa) and low overall precipitation amounts (666 mm), while MOD, MED and MP caves receive more precipitation than other coastal sites and summer droughts are less pronounced (Cfa). MP cave is also unusual on account of the relatively large difference between its surface MAAT (13.7 °C in 2014–2015 and 14.8 °C in 2018–2020) and a cave MAAT of 9.0 °C (1σ = 0.4) in 2012–2014. This is a consequence of cave morphology, i.e., descending chambers acting as a 'cold trap' [35]. Thus, it can be regarded as a transitional cave with climatic properties of both coastal and continental sites, as it has average summer temperatures higher than the continental sites, but also more precipitation than other coastal sites.

Table 1. Geographical positions of the caves, average cave air temperature, climate type of the cave region, monitoring and sampling periods and number of drip water samples. Colours of climate types correspond with the colours of the cave clusters in Figure 2. Monthly integrated drip water samples are given in bold and same locations have undergone systematic monitoring with monthly integrated rainwater sampling. Temperatures marked with (*) are averaged from [36] and [47] and with (**) are marked temperature ranges of deep vertical caves (see [48–51] for details).

Cave	Location	Acronym	Entrance			Clim. Type	Aver. Temp. (°C)	Monitoring/Sampling Period	Number of Drip Water Samples	Ref.
			Lat. (N)	Long. (E)	Alt. (m)					
Nova Grgosova	Samobor Hills	NG	45°49′	15°41′	239	Cfb	11.2	November 2014–November 2015	**36**	[25]
Lokvarka	Gorski Kotar	LOK	45°22′	14°46′	780	Cfb	7.5	November 2014–November 2015	**24**	[25]
Gornja Baraćeva	Kordun	GB	44°59′	15°43′	331.5	Cfb	10.2*	April 2013–July 2014	**24**	[36]
Donja Baraćeva	Kordun	DB	44°59′	15°43′	309.5	Cfb	10.1 *	April 2013–July 2014	**28**	[36]
Sirena	N Mt Velebit	SIR	44°46′	14°59′	1500	Df	1.7–3.6 **	2 August 2013	1	[39]
Lukina jama	N Mt Velebit	LUK	44°46′	15°02′	1450	Df	0–5 **	5 August 2013	1	[39]
Velebita Cave System	N Mt Velebit	VEL	44°45′	14°59′	1550	Df	3.5–5.5 **	31 July2012–4 August 2013 1 August 2013	18	[39]
Medvjeđa špilja	Lošinj Island	MED	44°36′	14°24′	17.5	Cfa	15	13 December 2009–10 January 2010	1	[41]
Špilja u Zubu Buljme	S Mt Velebit	ZB	44°22′	15°28′	1250	Df	4.0	July 2012–July 2013	**11**	[35]
Manita peć	S Mt Velebit	MP	44°19′	15°29′	570	Cfa	9.0	July 2012–July 2013	**32**	[35]

Table 1. *Cont.*

Cave	Location	Acronym	Entrance			Clim. Type	Aver. Temp. (°C)	Monitoring/Sampling Period	Number of Drip Water Samples	Ref.
			Lat. (N)	Long. (E)	Alt. (m)					
Donja Cerovačka	Lika	DC	44°16′	15°53′	630	Cfb	6.8	April 2013–January 2015	21	[36]
Modrič	Mt Velebit foothill	MOD	44°15′	15°32′	32	Cfa	16.6	June 2007–September 2007 July 2007–May 2008 November 2007–May 2008 June 2008–September 2010 November 2014–November 2018	3 1 1 54 93	[43] [37] [25]
Kraljičina spilja	Vis Island	KRA	44°04′	16°06′	70	Csa	14.1	19–20 May2010	1	[41]
Strašna peć	Dugi otok Island	SP	44°00′	15°02′	74	Csa	11.1	20 June–29 September 2010 July 2012–January 2014	2 36	[41] [35]
Špilja u Vrdolju	Brač Island	VRD	43°21′	16°36′	310	Csa	13.5	18–19 May2010	1	[41]
Velika špilja	Mljet Island	VML	42°46′	17°28′	90	Csa	14	20–22 February 2010	2	[41]
Mala špilja	Mljet Island	MML	42°46′	17°29′	60	Csa	13	20–22 February 2010	1	[41]

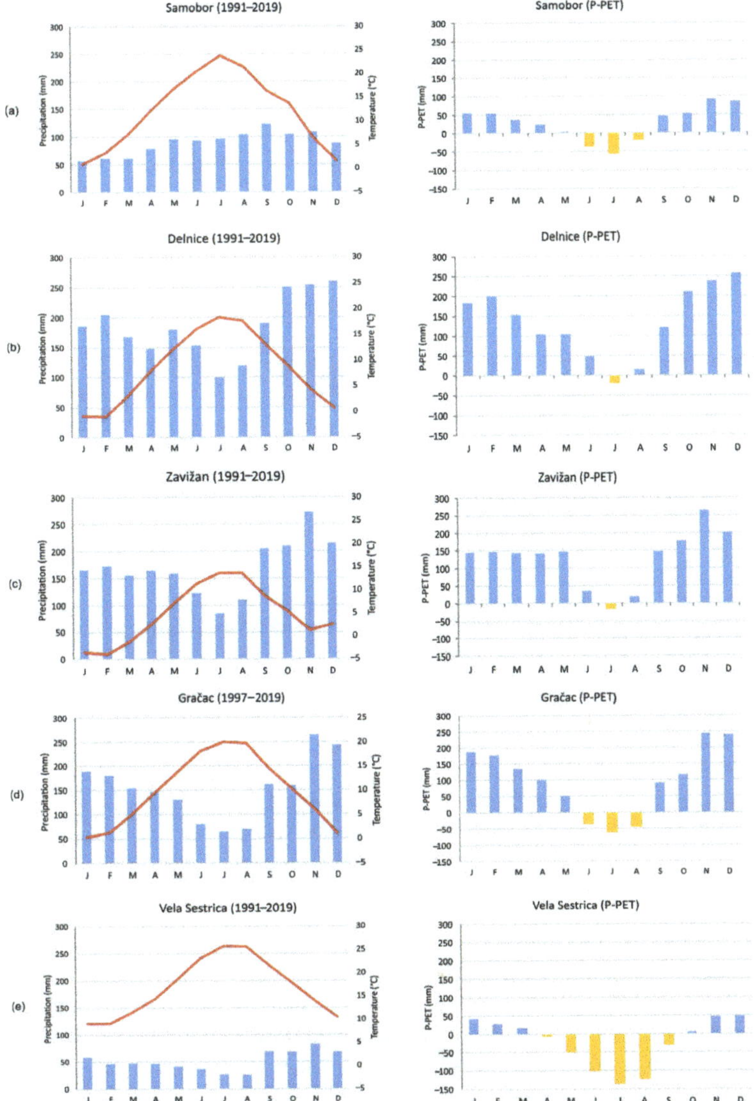

Figure 3. Climate data (left: air temperature and precipitation; right: difference between precipitation and potential evapotranspiration) for five characteristic regions: (**a**) Samobor station representing continental NG cave; (**b**) Delnice station representing continental settings with orographic effect on LOK cave; (**c**) Zavižan station for high-altitude conditions with a strong orographic effect (ZB, VEL, SIR, LUK); (**d**) Gračac station as representative for continental sites with a rain-shadow effect (DB, GB, DC caves); and (**e**) Vela Sestrica station representing conditions of low-altitude caves located on the islands and along the coast (MOD, SP, MED, VRD, KRA, MML, VML). Data obtained from the Croatian Meteorological and Hydrological Service [52]. Locations of the meteorological stations are given in Figure 1. Water balance (potential evapotranspiration) is calculated using the Thornthwaite evapotranspiration model [53,54].

3. Monitoring and Sampling Methodology

An overview of monitoring duration and number of samples is given in Table 1. Monitoring and sampling practice for the systematic long-term studies (NG, LOK, MOD, ZB, MP, SP, DB, GB, DC caves) was the same at all sites, and included recording meteorological parameters (air temperature and relative humidity) outside and inside the caves and measuring the isotopic composition (δ^2H and δ^{18}O) of rain and cave drip water [25,35,36]. Additionally, analyses of drip water elemental composition were performed at DB and GB sites [36] and in MOD cave (2012–2013, unpublished), while in NG, LOK, MOD, ZB, MP and SP caves drip intensity was recorded [25,35].

Microclimate data were measured at one-hour resolution by HOBO U23 Pro v2 Temperature/Relative Humidity Data Loggers. Loggers were mounted at 1.5–2.0 m above the cave floor and 2–3 cm off the wall to avoid recording the thermal signal of the bedrock, which is delayed, usually dampened [55] and could be misleading.

Monthly-integrated rainwater samples have been collected above ZB, MP, SP [35] NG, LOK and MOD caves [25] in plastic containers with funnel protected by PVC net to prevent clogging and a layer of paraffin oil to hamper the evaporation of collected rainwater, as this significantly effects water isotopic composition. During the study of MOD cave by Rudzka et al., [37], a Pluvimate® rain-gauge was used in order to collect rainwater and record its intensity. At the GB/DB and DC sites composite monthly samples of precipitation were collected by simple anti-evaporation system without an oil film, while at VEL, SIR and LUK sites the system of Gröning et al. [56] was employed [39]. Daily precipitation amount data were obtained from adjacent meteorological stations operated by the Croatian Meteorological and Hydrological Service (CMHS).

Monthly-integrated drip water samples were collected under the drip sites that had been feeding stalagmites predetermined for paleoclimate studies in NG, LOK, MOD, ZB, MP, SP, DB, GB, DC, as well as from several other drip points. In-cave water containers did not include protective paraffin oil since the cave air RH was 100%, as anticipated, so no evaporation was expected. It has also been empirically proven that equilibration of the collected water with the moisture of the cave atmosphere takes approximately three months [57], i.e., collected drip water samples would not have been compromised within one month. Regardless of the negative water balance during the summer season (especially pronounced in coastal region, e.g., Vela Sestrica station in Figure 3), it must be emphasized that practically all our drip sites had continuous discharge all year round. Research in the high-mountain Velebita Cave System (VEL), which is a 1026 m deep vertical cave, focused on the isotopic composition variability along a vertical profile and its comparison with two other pits (LUK and SIR), with local precipitation and with some remote spring waters. Thus, along the VEL profile, 1 to 6 samples were collected daily during five days of caving, while in LUK and SIR only one sample of water in each cave was collected [39]. In island caves (SP, MED, VRD, KRA, MML, VML) the single samples were also collected per drip site, with drip water accumulating from 2 to 102 days [40,41]. All drip water samples were taken from the exact sites from where speleothems have been collected for paleoenvironmental reconstruction.

Modern calcite was collected on pre-cleaned glass plates (10 × 10 cm) in MP caves, on plastic funnels and glass plates in SP cave [35] and on a nylon sheet in MOD cave [43]. Farming of modern calcite in DB and GB caves was performed by placing light bulbs under active drip sites, imitating the rounded stalagmite surface [36]. We avoided sampling of the youngest parts of the active stalagmites since with the artificial material we were sure about the period of deposition. Analytical methods related to stable isotope measurements used in Rudzka et al., [37], Lončar [40], Surić et al., [25,35], Czuppon et al., [36] and Paar et al. [39] are given in detail in Appendix A.

Valuable information on aquifer architecture can be obtained from the relationship between precipitation and cave discharge [58], so NG, LOK, MOD, ZB, MP and SP caves were also equipped with automated acoustic Stalagmate® drip counters [59] to record discharge dynamics.

4. Results and Discussion

A statistical summary of the $\delta^{18}O$ and δ^2H data from the majority of analyzed water samples is given in Tables 2 and 3. Systematic monitoring and sampling of precipitation revealed nine local meteoric water lines (LMWL) (linear regression between precipitation $\delta^{18}O$-δ^2H monthly values) (Figure 4) along which corresponding cave drip water $\delta^{18}O$-δ^2H values are distributed according to different site-specific environmentally dependent patterns. The relationship between the isotopic composition of precipitation and cave drip water is presented in Figures 5–7, and spatiotemporal patterns of precipitation isotopic signal transmitted to the cave drip water are discussed in the following sections.

Table 2. Minimum, maximum and amplitude of precipitation and drip water $\delta^{18}O$ and δ^2H from NG, LOK, MP, GB, DB, ZB, MP, DC, MOD and SP caves for the periods given in Table 1. Sites are arranged by latitude. All the values are given in ‰ with respect to Vienna Standard Mean Ocean Water VSMOW.

| | Precipitation | | | | | | | Drip Water | | | | | |
| | Maximum | | Minimum | | Amplitude | | | Maximum | | Minimum | | Amplitude | |
Cave	$\delta^{18}O$	δ^2H	$\delta^{18}O$	δ^2H	$\delta^{18}O$	δ^2H	Drip Site	$\delta^{18}O$	δ^2H	$\delta^{18}O$	δ^2H	$\delta^{18}O$	δ^2H
NG	−4.6	−24.6	−13.5	−93.5	8.9	68.9	NG-1W	−9.5	−63.1	−10.1	−65.8	0.6	2.7
							NG-2W	−9.6	−63.3	−10.0	−65.6	0.4	2.3
							NG-3W	−9.6	−63.1	−9.9	−64.1	0.3	1.0
LOK	−5.3	−29.0	−13.9	−96.8	8.6	67.8	LOK-1W	−7.9	−45.9	−9.7	−58.5	1.3	12.6
							LOK-2W	−8.5	−50.2	−8.8	−52.1	0.3	1.9
GB/DB	−5.4	−34.6	−12.9	−86.8	7.5	52.2	GB-1	−9.5	−64.0	−11.0	−74.2	1.5	10.2
							GB-2	−10.2	−70.7	−11.8	−82.0	1.6	11.3
							DB-1	−10.0	−70.7	−11.1	−77.1	1.1	6.4
							DB-2	−10.0	−71.5	−10.8	−74.2	0.8	2.7
ZB	−5.9	−30.9	−13.4	−93.2	7.5	62.3	ZBW	−7.4	−41.2	−8.2	−48.4	0.8	7.2
MP	−3.4	−16.3	−9.4	−58.9	6.0	42.6	MP-1W	−6.3	−36.1	−7.9	−48.4	1.6	12.3
							MP-2W	−6.6	−36.3	−7.5	−43.1	0.9	6.8
							MP-3W	−6.4	−33.3	−7.7	−47.9	1.3	14.6
DC	−5.3	−32.3	−10.7	−73.9	5.4	41.6	DC-2	−7.5	−46.0	−8.4	−51.8	0.9	5.8
MOD	−3.2	−17.0	−8.6	−57.1	5.4	40.1	MOD-21W	−5.7	−33.6	−6.8	−41.7	1.1	8.1
							MOD-22W	−3.9	−17.0	−6.0	−35.3	2.1	18.3
	−1.2	−11.1	−9.8	−60.9	8.6	49.8	MOD-31W	−5.6	−34.1	−6.3	−38.6	0.7	4.5
							MOD-32W	−5.9	−35.7	−6.4	−38.8	0.5	3.1
SP	−3.8	−23.2	−8.3	−50.6	4.5	27.4	SP1W	−6.0	−32.3	−6.8	−35.8	0.8	3.5
							SPNW	−6.2	−33.2	−6.7	−38.8	0.5	5.6

4.1. Moisture Sources Affecting Isotopic Composition

The aim of many speleothem-based studies is to assess past dynamics of atmospheric water vapor masses e.g., [60–63], as their shifts have governed associated precipitation, temperature and vegetation distribution. Of special interest are bordering regions such as the one between continental European and Mediterranean influences, and in that sense the Croatian region could play an important role, since this spatially small area rich in geographical and climatic variety is presently characterized by complex patterns of precipitation isotopes [26,27,32,35]. Figure 4 shows regression lines of nine sets of precipitation data. Since some of the lines are based on one year of data, the R^2 is given to depict statistically significant correlations. The most obvious characteristic of this set of LMWLs is their departure from the Eastern Mediterranean meteoric water line (EMMWL: $\delta^2H = 8 \times \delta^{18}O + 22$ [64]) which represents environments characterized by substantial evaporation under conditions of a large humidity deficit [65]. This is apparently not the case in our studied region. The LMWLs are generally clustered along the Global meteoric water line (GMWL: $\delta^2H = 8 \times \delta^{18}O + 10$ [66]) and the western Mediterranean meteoric water line (WMMWL: $\delta^2H = 8 \times \delta^{18}O + 13.7$ [67]), which marks the transitional zone with mixing influences of warm eastern Mediterranean and cool North Atlantic air masses [67]. Another characteristic of these water lines are the slopes, which are systematically lower (from 6.6 to 7.8) than that of the GMWL, with a distinct difference between northern continental stations (closer to the GMWL slope), and maritime sites with lower values. The reason for this is secondary evaporation which occurs as rain falls through a dry air column above the ground and leads to kinetic fractionation [68]. This is a typical situation during the coastal dry and hot summers (Figure 3e), while the uniform annual distribution of the precipitation partially suppresses that effect in continental sites (Figure 3a,b). However, it is obvious that all caves receive mixed precipitation from the Atlantic and Mediterranean regions, but with different seasonal patterns.

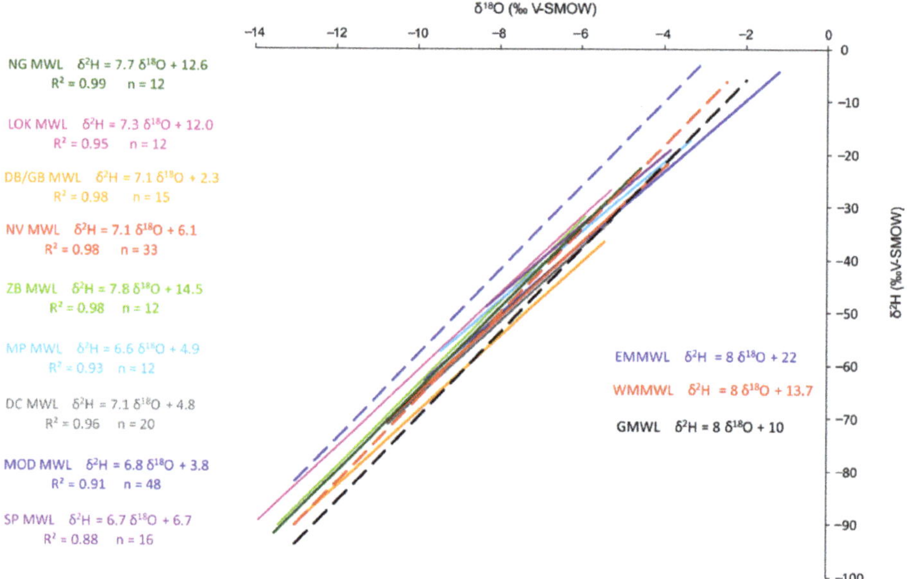

Figure 4. Local meteoric water lines (LMWL) of nine studied locations (NG, LOK, GB/DB, NV, ZB, MP, DC, MOD, SP) plotted alongside global (GMWL: $\delta^2H = 8 \times \delta^{18}O + 10$ [66]), eastern Mediterranean (EMMWL: $\delta^2H = 8 \times \delta^{18}O + 22$ [64]), and western Mediterranean meteoric water line (WMMWL: $\delta^2H = 8 \times \delta^{18}O + 13.7$ [67]) – dashed lines. NV relates to 'North Velebit', as the rainwater was collected at three locations in VEL, SIR and LUK caves' vicinity. For clarity, single-sample data points are removed, but they are visible in Figures 5–7 for each particular LMWL. Equations are sorted by the latitude of the site.

4.2. Precipitation-Discharge Relationship and Drip Water Homogenization

When reconstructing multi-annual environmental changes using slow-growing stalagmites, one of the first issues to consider when interpreting particular drip water isotopic signals is the homogenization of the meteoric water during its transition through the epikarst towards the drip site, to ensure that the recorded speleothem $\delta^{18}O$ signal is not seasonally biased. Homogenization of the drip water, i.e., removal of the seasonal signal of precipitation, had been initially taken as a rule [69], but it has since been realized that it must be considered as a site-specific process which is not always achieved, sometimes not even within the same cave [9,25,70–72]. As presented in Table 2, seasonal variations of rainwater $\delta^{18}O$ (amplitudes from 4.5‰ to 8.9‰) and δ^2H (amplitudes from 27.4‰ to 68.9‰) show some geographical/climate related distribution, being smaller in southern and coastal regions, while the northern and continental sites receive precipitation with a wider isotopic range. On the other hand, the smaller amplitude of drip water $\delta^{18}O$ (from 0.3‰ to 2.1‰) and $\delta^{2}H$ (from 1.0‰ to 18.3‰) shows significant attenuation of the seasonal rainwater signal, regardless of geographical position.

Groundwater homogenization strongly depends on karst aquifer architecture, which also plays an important role in controlling drip discharge dynamics [73]. Infiltration-discharge relationships were observed in six caves and drip intensities recorded in caves with multiple drip-loggers (NG, LOK, MP, MOD, SP), confirming heterogeneity of the karst aquifers and their typical triple porosity nature with: (i) slow diffuse matrix flow, (ii) fracture and fissure flow and iii) fast preferential flow through solution-enhanced conduits [74,75]. Discharge with significant variability and coherency with surface rain events can tentatively point to the lower degree of homogenization, i.e., partially preserved seasonal signal, so drip water $\delta^{18}O$-$\delta^{2}H$ pairs plotted in $\delta^{18}O$-$\delta^{2}H$ space appear scattered along the precipitation regression line. For example, according to concurrent daily rainfall totals and drip-logger records, MP-1 and MP-3 (Figure 5) demonstrated fast responses to surface rain events through epikarst fissures [25]. Similarly, MOD-22 also showed some seasonal variability (Figure 5), but is strongly out of phase with rainfall values, with the highest values during the winter. Apparently, this is a consequence of 'piston-flow' when newly recharged rainwater pushes through previously stored and partly mixed water [37]. On the other hand, densely clustered drip water values like those of MOD-31 and MOD-32 (Figure 5) coincide with very stable discharge regimes, practically unresponsive both to droughts and massive rain events [25]. Still, the peculiarity of karst systems is evident in LOK and NG caves. Drip sites LOK-1 and LOK-2 show evidence of typical fracture-flow or 'seasonal drip' regimes according to the classification by Smart and Friederich [76] and Baker et al., [58], with practically identical drip records [25]. However, LOK-1 preserves an attenuated seasonal signal which is completely dampened in LOK-2 (Figure 6). The opposite situation is observed in NG cave; drip water isotopic compositions of NG-2 and NG-3 are almost the same with minimal amplitudes (Figure 6) in spite of substantially different hydrological behaviour. NG-3, with typical matrix porosity, sustains stable discharge throughout the year, while NG-2 probably has a more complex route that includes 'overflow' storage reservoirs [3] expressed as fast response to major rain events and subsequent long-term decrease in discharge [25].

Table 3. Amount-weighted annual mean of precipitation and drip water $\delta^{18}O$, $\delta^{2}H$ and *d*-excess. For the drip sites not equipped with drip loggers, only average ($\pm 1\sigma$) values are given. Sites are arranged by latitude. All the values are given in ‰ with respect to VSMOW.

Cave	$\delta^{18}O_{prec}$	$\delta^{2}H_{prec}$	$\delta^{18}O_{drip}$	$\delta^{2}H_{drip}$	$\Delta\delta^{18}O_{prec-drip}$	$\Delta\delta^{2}H_{prec-drip}$	*d*-excess-Prec.	*d*-excess-Drip
NG	−9.1	−57.1	−9.8 −9.8	−64.3 −63.5	0.7 0.7	7.2 6.4	15.6	14.3 14.9
LOK	−8.8	−51.8	−8.9 −8.6	−52.8 −50.9	0.1 −0.2	1.0 −0.9	18.6	18.5 18.0
GB-DB	−9.5	−64.9	−10.7 ± 0.3 −10.6 ± 0.2 −10.3 ± 0.5 −10.7 ± 0.5	−73.7 ± 2.1 −72.7 ± 0.7 −70.2 ± 3.0 −73.7 ± 3.5	1.2 ± 0.3 1.1 ± 0.2 0.8 ± 0.5 1.2 ± 0.5	8.8 ± 2.1 7.8 ± 0.7 5.3 ± 3.0 8.8 ± 3.5	11.0	12.0
VEL-S [1] VEL-O [2]	−9.5 −9.4	−61.9 −60.9	−9.6 ± 0.1 [3]	−61.9 ± 0.7 [3]	0.1 ± 0.1 0.2 ± 0.1	0.0 ± 0.7 1.0 ± 0.7	13.7 13.1	15.0 ± 0.5
ZB	−9.9	−62.2	−7.8 ± 0.3	−44.6 ± 2.5	−1.2 ± 0.3	−17.6 ± 2.5	16.6	16.3 ± 0.6
MP	−7.5	−44.3	−7.3	−42.0	−0.2	−2.3	15.5	16.2
DC	−8.6−−7.5 [4]	−56.5−−48.1 [4]	−8.0 ± 0.3	−49.0 ± 1.7	−0.6−0.5	−7.5−0.9	12.0–13.0	15.0
MOD	−5.5	−33.3	−6.0 −6.1	−36.5 −37.2	0.5 0.6	3.2 3.9	10.3	11.3 11.8
SP	−6.7	−38.3	−6.5	−34.4	−0.2	−3.9	15.2	17.3

[1] precipitation collected at location Siča; [2] precipitation collected at location Oltari; [3] cave water was collected only during five days of caving in August 2012; [4] different values depending on the considered period.

4.3. Isotopic Composition of Drip Water

The relationship between rain and drip water δ^2H-$\delta^{18}O$ pairs of coastal and insular sites is shown in Figure 5. MOD, MP and SP caves are presented by at least one-year series of rain and drip water monthly—integrated samples and associated LMWLs, while MED, VRD, KRA, VML and MML are represented by data from one–two drip water samples. Just the latter are plotted, and further analyses of these sites are unfeasible without extended monitoring. When compared to continental sites, the series from MP, SP and 2008–2010 MOD rainwater data have a relatively narrow range (Table 2) due to the moderate maritime effect on temperature [68], with the exception of the longest MOD data set (2014–2018) which encompasses several extreme meteorological events with the most anomalous values. The similarity of LMWL slopes points to the same water vapor sources and infiltration conditions, but the aquifer architecture subsequently assumes the lead role. In the case of MOD cave, we can distinguish three different drip regimes and corresponding isotopic compositions (Table 2); MOD-21W and MOD-22W had lower and higher variability, respectively, while MOD-31W and MOD-32W are characterized by strong drip water homogenization [25]. Therefore, we took these sites as representative since that is the preferable hydrologic behavior for the best climate record in speleothems. The amount-weighted annual mean of precipitation $\delta^{18}O$ is 0.5–0.6‰ higher than the amount-weighted annual mean of drip water $\delta^{18}O$ (Table 3), meaning that effective infiltration and accumulation within aquifer occurs during the winter period. As for MP and SP caves, they both had amount-weighted annual means of precipitation $\delta^{18}O$ 0.2‰ lower in comparison to that of drip water (Table 3). These caves are very geographically different, insular SP at 70 m a.s.l. and mountain MP at 570 m a.s.l., but they have similar morphologies, i.e., descending chambers which maintain significantly lower temperatures. In SP cave, the MAAT is 11.1 °C, while surface MAAT is 16.4 °C, and in the case of MP, cave MAAT is 9.0 °C and surface MAAT is 13.7 °C [35]. This difference of 5.0 ± 0.3 °C complicates the classification of these caves in terms of climatic controls on the isotopic composition of drip water as discussed in Baker et al. [77].

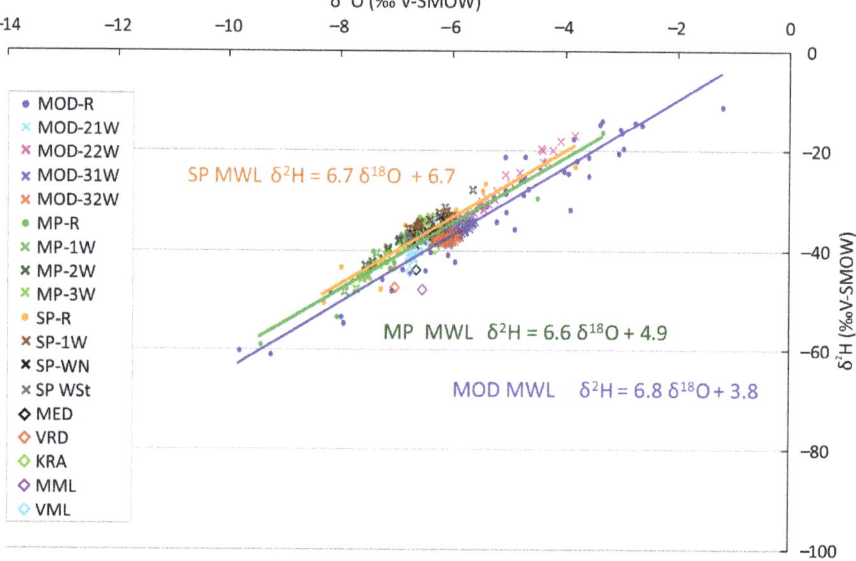

Figure 5. δ^2H-$\delta^{18}O$ relationship of monthly integrated drip water (crosses) and rainwater samples (dots) with associated local meteoric waterlines in SP, MP and MOD caves, and of single drip water samples from low-altitude coastal caves MED, VRD, KRA, VML and MML (diamonds).

The relationship between precipitation and drip water isotopic composition of continental sites NG, LOK, GB, DB and DC is given in Figure 6. δ^2H and $\delta^{18}O$ have wide ranges due to the strong seasonal temperature variations [68], among which DC's smallest amplitude (Table 2) reflects its southernmost position (Figure 1). Drip water isotopic values of NG, GB and DB sites suggest larger contributions of winter precipitation, with a difference between amount-weighted annual mean precipitation and average drip water $\delta^{18}O$ of 0.7‰ in NG cave, 0.8–1.2‰ in GB and 1.1–1.2‰ in DB cave (Table 3), at least for the monitoring period [36]. Caves with lower temperatures (LOK and DC) showed a different pattern; their drip water mean $\delta^{18}O$ values were distributed around the amount-weighted annual mean of precipitation $\delta^{18}O$ (Table 3), with indistinguishable preferential recharge period.

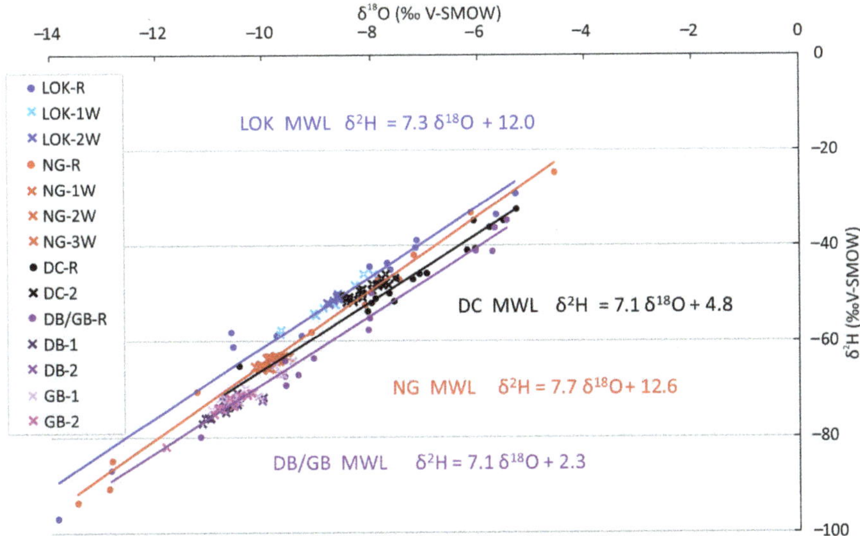

Figure 6. δ^2H-$\delta^{18}O$ relationship of monthly integrated drip water (crosses) and rainwater samples (dots) with associated local meteoric waterlines in mid-altitude continental caves NG, LOK, DB, GB, DC.

The hydrology of the high-mountain caves is represented by two LMWLs (Figure 7). North Velebit caves (VEL, LUK, SIR) located at 1450–1550 m a.s.l. are in fact vertical shafts; LUK and SIR caves are characterized by perennial snow and ice accumulation in the upper portions [49–51,78], while VEL cave is ice-free due to its morphology, i.e., horizontal entrances [49]. According to the drip water samples from VEL cave, these aquifers are predominantly fed by September–May precipitation [39]. Average annual duration of snow cover ($\geq$ 1 cm) in that region is 165 days (Zavižan station 1981–2010, CMHS, 2020) and most of the recharge is meltwater, hence the grouping of drip water δ^2H-$\delta^{18}O$ data along the lowermost part of the NV MWL (Figure 7). The difference between the amount-weighted annual mean precipitation and average drip water $\delta^{18}O$ is 0.1–0.2‰ (Table 3, Figure 8), although Figure 7 would suggest a larger difference.

ZB cave is at a lower altitude (1250 m a.s.l.) and is not an ice cave; snow cover lasts considerably shorter than in the North Velebit sites. Its LMWL clearly shows partly evaporative conditions at the water-vapor source and minimal secondary evaporation during precipitation (Figure 7), consistent with its geographic location. Drip water was relatively uniformly discharged and homogenized [35]. However, the isotopic composition demonstrates significant departure towards the summer values (Figure 7), so the amount-weighted annual mean of precipitation $\delta^{18}O$ was 1.2 ± 0.3‰ lower than the annual mean of drip water $\delta^{18}O$.

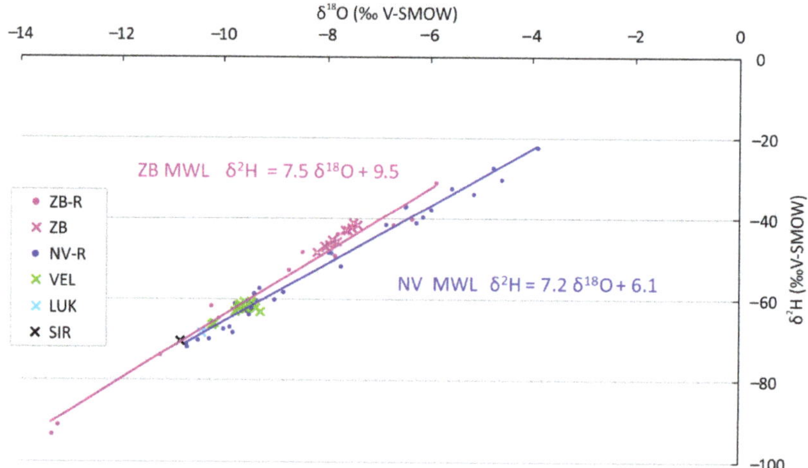

Figure 7. δ^2H-$\delta^{18}O$ relationship of monthly integrated drip water (crosses) and rainwater samples (dots) with associated local meteoric water lines in high-altitude mountain—caves ZB, VEL, SIR, LUK. NV relates to 'North Velebit', as the rainwater was collected at three locations in VEL, SIR and LUK caves vicinity.

Recently, Baker et al., [77] conducted a global analysis of rain and drip water data from 39 caves on five continents and demonstrated that the extent to which drip water $\delta^{18}O$ is representative of amount-weighted precipitation $\delta^{18}O$ is related to MAAT and annual precipitation, which determines the extent to which the isotopic signal is additionally altered by soil and karst processes. It was concluded that caves with a MAAT <10 °C show minimal differences between drip water $\delta^{18}O$ and amount-weighted precipitation $\delta^{18}O$; Croatian 'cold' caves LUK, VEL, SIR, DC and LOK generally fit into that classification (Figure 9), i.e., their $\Delta\delta^{18}O_{prec-drip}$ vary around 0‰. The peculiar exception is ZB cave with a large drip water $\delta^{18}O$ departure towards summer values ($\Delta\delta^{18}O_{prec-drip} = -1.2 \pm 0.3‰$) (Figure 8). Given the relatively uniform discharge (checked by manual drip counting during each visit [35]) and homogenized drip water (attenuation of rainfall $\delta^{18}O$ amplitude of 7.5‰ to the drip water $\delta^{18}O$ range 0.8‰, Table 3) it is apparent that the isotopic signal of any individual recharge events is buffered, giving the credible drip water average. Such a significant $\delta^{18}O$ increase may be ascribed to evaporative fractionation, which is quite unrealistic in a high-mountain region with a MAAT of 4.0 °C, but on the other hand the constant in-cave atmospheric conditions in ZB has been occasionally disturbed and RH has decreased during dry bora wind episodes [35].

Other studied Croatian regions would fit into the class with MAAT between 10 °C and 16 °C (we included here also MOD region with MAAT 16.6 °C, as a marginal case) in which the drip water $\delta^{18}O$ may be affected by evaporative fractionation of the water in the soil and epikarst or by selective recharge, resulting in $|\Delta\delta^{18}O_{prec-drip}| > 0.3‰$ [77]. Such values are recorded in MOD, NG, DB and GB caves, but not in the two littoral caves, MP and SP, which are characterized with $|\Delta\delta^{18}O_{prec-drip}| < 0.3‰$, as if they were 'cold' caves. In particular, the amount-weighted annual means of precipitation $\delta^{18}O$ are lower by 0.2‰ in comparison to that of drip waters which is a consequence of the combined influence: of (i) evaporative fractionation due to relatively high MAAT and especially summer temperatures; (ii) selective recharge during the autumn and winter seasons; and (iii) the decrease of cave MAAT in relation to surface MAAT by 5.0 ± 0.3 °C.

Apart from the occasionally necessary drip-specific interpretation, and morphologically driven exceptions, it must be considered that during glacial periods, today's 'warm' caves assumed cold-cave characteristics. Thus, within a single speleothem which covers one or more glacial-interglacial shifts, different climate control on $\Delta\delta^{18}O_{prec-drip}$ would be experienced [77].

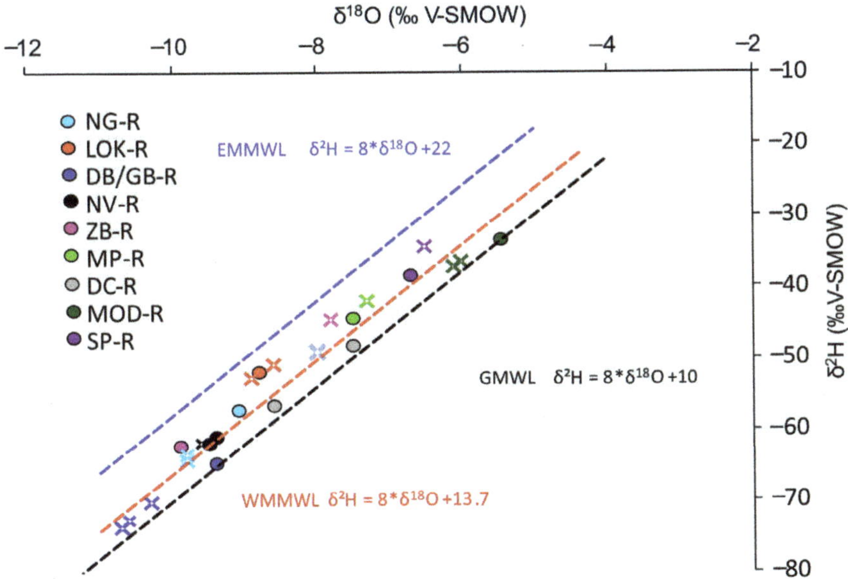

Figure 8. Amount-weighted annual mean of precipitation (circles) and drip water $\delta^{18}O$ and δ^2H (crosses) in relation to the global (GMWL: $\delta^2H = 8 \times \delta^{18}O + 10$ [66]), eastern Mediterranean (EMMWL: $\delta^2H = 8 \times \delta^{18}O + 22$ [64]), and western Mediterranean meteoric water line (WMMWL: $\delta^2H = 8 \times \delta^{18}O + 13.7$ [67]). The LUK-VEL-SRI values represent precipitation collected at two neighboring sites (Siča and Oltari [39]) during 2012–2013 for 17 months, but with several gaps, so only tentative values are given (calculated as averages of several different intervals). For the unlogged drip sites (ZB, GB, DB, DC), we used annual averages instead of amount-weighted $\delta^{18}O$ and δ^2H values. Sites are arranged by latitude in the legend box.

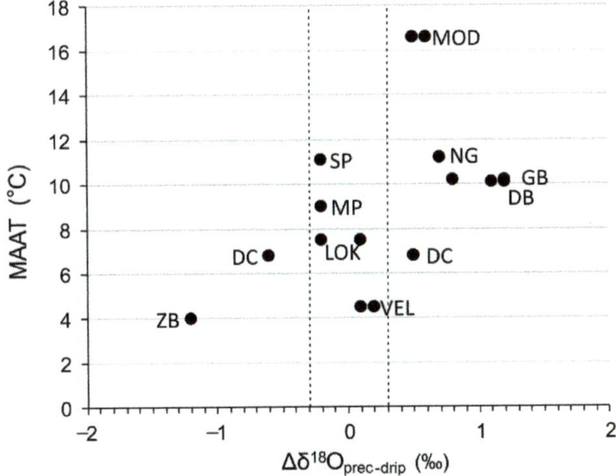

Figure 9. Relationship between $\Delta\delta^{18}O_{prec-drip}$ and mean annual air temperature (MAAT). Vertical lines show the 0.3‰ criterion for determining the significant difference between amount-weighted annual means of precipitation $\delta^{18}O$ and drip water $\delta^{18}O$ [77].

4.4. Preservation of Altitude and Latitude Effects in Drip Water

Due to the upward deflection by the mountain barrier, air masses decompress and cool adiabatically, so the rainout is enhanced [79], and due to the temperature decrease with elevation, the 'altitude effect' is characterized by a decrease in $\delta^{18}O$ values. When considering precipitation, an average gradient ($\Delta\delta^{18}O$/100 m) of −0.2‰ is estimated globally [80], but since topography strongly influences isotopic composition, especially if the relief is high [79], local gradients are even higher. When comparing precipitation and drip water isotopic composition, infiltration elevation must be taken into account since the altitude of the catchment area can be substantially higher than the elevation of the drip water collection site, delivering more depleted water. Our observed profile extends from MOD cave at 32 m a.s.l. via MP cave (570 m a.s.l.) to ZB at 1250 m a.s.l. across a horizontal distance of 13.6 km, and according to the amount-weighted mean annual precipitation $\delta^{18}O$ (−5.5‰, −7.5‰ and −9.9‰, respectively), the local $\Delta\delta^{18}O$/100 m gradient is −0.36‰ (Figure 10). As for the infiltration elevation of these three caves, their recharge areas are limited by relatively thin overburden and positions near the summits, so drip water does not carry a depleted isotopic signal from much higher elevations. The observed drip water $\Delta\delta^{18}O$/100 m gradient is −0.15‰ (Figure 10), comparable to steep north Italian Alps where two transects had gradient of −0.15‰ and −0.08‰ [81]. It is also similar to a steep coastal area on Vancouver Island where the drip water gradient was −0.16‰ [72] and the western flank of Mount-Lebanon where it was −0.2‰ [82]. However, MP and ZB sites, as mentioned above, may not be the best representatives for general conclusions on the altitudinal effect on drip water due to their specific settings, but it is indicative that the drip water gradient is attenuated according to precipitation gradient.

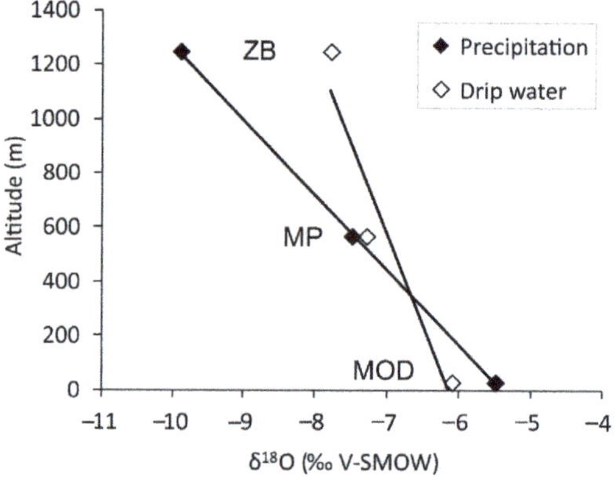

Figure 10. Comparison of altitudinal gradient in precipitation and drip water on steep Mt Velebit seaward slope based on MOD (32 m a.s.l.), MP (570 m a.s.l.) and ZB (1250 m a.s.l.) site values $\delta^{18}O$.

The latitude effect on precipitation $\delta^{18}O$ in mid-latitude regions is about −0.5‰ $\delta^{18}O$ per degree of latitude [77]. In the studied Croatian sites, the gradient is apparent (Figure 8), but the absolute value is undetectable as it is partly overprinted by the altitude effect. The same situation is evident in the drip waters.

4.5. Isotope Fractionation during Carbonate Formation

Based on the previous sections, it is evident that the drip water and its stable isotope composition have great significance in preserving and inheriting climatic information that can be transferred to the speleothem. However, this preservation strongly depends on the fraction which took place during the carbonate formation. Therefore, it is essential to establish the equation which describes the isotope fraction during the speleothem formation (i.e., carbonate precipitation) at each cave site.

The stable isotope composition of modern calcite and concomitant drip water together with the cave air temperature can be utilized to select the equations which best express isotope fractionation. Among the caves reviewed in this study, modern carbonate precipitates were collected at MOD, MP, DB, GB and SP and the related data are summarized in Table 4.

Table 4. Oxygen isotope composition of modern calcite and concomitant drip water and measured (T_m) and calculated (T_c) cave air temperature. T_c values closer to the T_m are emboldened. Note that for calculations, the $\delta^{18}O_{calcite}$ value was converted from VPDB (Vienna Pee Dee Belemnite) to VSMOW using the expression $\delta^{18}O_{VSMOW} = 1.03091 \times \delta^{18}O_{VPDB} + 30.91$ [83]. Temperatures are calculated according to Tremaine et al., [84] and Daëron et al., [85]. The uncertainties of the $\delta^{18}O$ analyses (both for calcite and water) is generally 0.1‰ (see Appendix A).

Cave	Sample ID	$\delta^{18}O_{calcite}$ (‰ VPDB)	$\delta^{18}O_{drip\ water}$ (‰ VSMOW)	T_m (°C)	T_c (°C) Tremaine et al. (2011)	T_c (°C) Daëron et al. (2019)
MOD	MOD 8	−5.5	−5.8	15.6	**17.7**	19.4
	MOD 9	−5.3	−5.8	15.6	**16.5**	18.3
MP	MPC3	−5.4	−7.2	9.2	**9.8**	12.1
	MPC4	−5.5	−7.2	9.2	**10.4**	12.6
	MPC5	−5.4	−7.2	9.2	**10.1**	12.3
	MPC6	−5.2	−7.2	9.2	**9.1**	11.5
SP	SPC2	−4.3	−6.6	11.1	7.5	**9.9**
	SPC3	−4.4	−6.6	11.1	8.2	**10.5**
	SP1WC	−5.0	−6.6	10.3	**11.0**	13.2
	SP1WC	−4.7	−6.8	10.6	8.3	**10.7**
	SPNC	−4.8	−6.5	10.3	**10.6**	12.8
DB	DB1	−8.1	−10.7	9.9	6.2	**8.7**
	DB2	−8.9	−10.6	11.1	10.5	**12.7**
GB	GB1	−8.7	−10.3	9.7	**10.9**	13.1
	GB2	−8.4	−10.7	10.9	7.6	**10.1**

In Figure 11 the calculated values are plotted along with equations based on theoretical, experimental and empirical investigations [11,84–87]. Most of the data are scattering along the equations line obtained from empirical observation by Tremaine et al. [84] ($1000\ln\alpha = 16.1 \times 10^3/T - 24.6$), and Daëron et al. [86] ($1000\ln\alpha = 17.57 \times 10^3/T - 29.13$); the latter is almost identical to the equation determined by Coplen (2007). Therefore, utilizing these equations and the measured stable isotope composition, the temperature can be calculated (T_c) and compared with those observed in the cave (T_m). There are also some data (e.g., MOD) which plot between equations by Tremaine et al., [84] and Friedmann and O'Neil [86] ($1000\ln\alpha = 2.78 \times 10^6/T^2 - 2.89$). Interestingly, even in the same cave system (e.g., DB/GB, SP) distinct isotope fraction can take place, drawing attention to the importance of studying in-situ calcite precipitation for all sites from where speleothems are taken for investigation in order to reconstruct past climate and environmental changes [36].

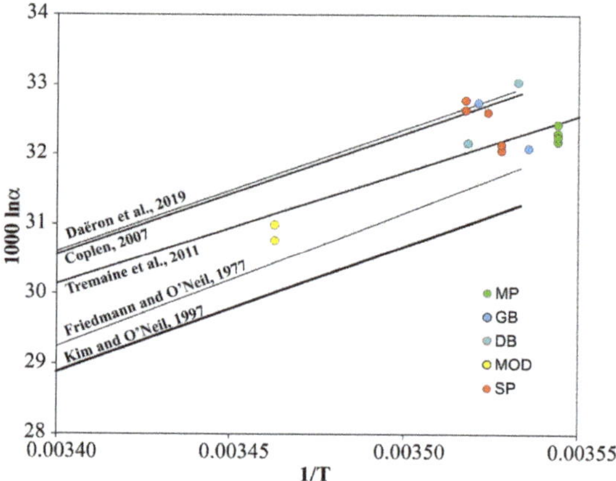

Figure 11. Calcite-water 1000·lnα values vs. 1/T (in K) measured for modern carbonates samples from MP, DB, GB, MOD and SP caves and selected published curves [11,84–87]. Calcite/water fractionation factor $\alpha = (1000 + \delta^{18}O_{calcite})/(1000 + \delta^{18}O_{water})$.

5. Conclusions

We revisited the stable isotope composition of precipitation and drip water from 17 Croatian caves with the intention of characterizing the main features essential for interpretation of speleothem isotope records in local and regional paleoenvironmental studies. To summarize:

- Nine constructed local meteoric water lines generally show influence of both Atlantic and Western Mediterranean vapor masses, and individually clearly depict particular regions with enhanced secondary evaporation.

- Drip waters in mountain and continental caves with MAAT < 10 °C drip water $\delta^{18}O$ mostly resemble the amount-weighted annual mean $\delta^{18}O$ of precipitation. These are the caves where the soil and epikarst evaporation has decreased, and groundwater is well mixed with a dampening of individual extreme events. Thus, in near-equilibrium calcite crystallization, speleothem $\delta^{18}O$ variations most faithfully reflect past meteoric precipitation and air temperature.

- In warmer caves, drip water $\delta^{18}O$ is usually more negative than amount-weighted annual mean precipitation $\delta^{18}O$ due to the uneven seasonal recharge and near-surface evaporation. However, when interpreting the interglacial-glacial transition it must be taken into consideration that MAATs have decreased to those of today's 'cold' caves and the isotopic signal probably directly reflects recharge.

- The precipitation altitude effect on seaward steep coastal mountains is higher than globally estimated, enhanced by sudden change from Mediterranean climate (Csa and Csb) to mountain (Df) climate. In drip water, the altitudinal gradient is also present, but less expressed.

- The latitude effect in drip water is present, but often overprinted by the altitudinal effect.

Among all derived conclusions, this is the first and foremost guiding principle against which all actions need to be conducted in forthcoming studies.

- This spatially small, but geographically very diverse area does not allow generalization; frequent exceptions of anticipated settings require thorough monitoring of surface and cave meteorological background since cave morphology (ascending/descending passages) may alter the cave temperature from those equal to the surface MAAT. Systematic analyses of drip water geochemistry may also reveal potentially different infiltration elevations, while monitoring of drip intensity provides information on aquifer architecture responsible for the water homogenization. Preferably, all of these steps should precede speleothem collection in order to avoid their oversampling.

Author Contributions: Conceptualization, methodology and investigation, M.S., R.L., N.L., G.C. and N.B.; formal analysis, R.N.D. and P.B.; visualization, M.S., N.B., R.L. and G.C.; writing—original draft preparation, M.S., R.L. and G.C.; writing—review and editing, M.S., G.C., R.L., N.B., P.B. and R.N.D. All authors have read and agreed to the published version of the manuscript.

Funding: This research received no external funding. Data have been obtained during the projects *Reconstruction of the regional paleoclimate change—speleothem records from the North Dalmatia (Croatia)* (60200) funded by the University of Zadar, *Reconstruction of the Quaternary environment in Croatia using isotope methods* (HRZZ-IP-2013-11-1623) funded by the Croatian Science Foundation, and *Research of geomorphological-geological conditions of karst and hydro-morphological peculiarities on selected localities of the Dinaric karst in Croatia* (20284703) funded by the University of Zagreb. N.L was granted with Unity Trough Knowledge Fund Grant (71/10) and G. C. was granted by Hungarian Scientific Research Fund (PD 121387) and János Bolyai Research Scholarship of the Hungarian Academy of Sciences.

Acknowledgments: Meteorological data were obtained on request (free of charge) from the Croatian Meteorological and Hydrological Service. Two anonymous reviewers and Guest Editor are gratefully acknowledged for the constructive suggestions that improved the manuscript.

Conflicts of Interest: The authors declare no conflict of interest. The funders had no role in the design of the study; in the collection, analyses, or interpretation of data; in the writing of the manuscript, or in the decision to publish the results.

Appendix A

Stable isotope composition of rain and drip water from NG, LOK, ZB, MP, SP and MOD (2014–2018) was carried out on ~2 mL subsamples of filtered sample water using a Picarro L2120 cavity ring-down isotope analyzer at the School of Geography, The University of Melbourne. Sample isotopic ratios were normalized to the international VSMOW scale using a four point calibration using two international (GISP2, VSMOW) and two in-house standards (WOOLIES, LAKE) of known composition. The results from the WOOLIES standards were used to determine the analytical uncertainty (1σ) which was $\leq 0.1‰$ for $\delta^{18}O$ and $\leq 0.3‰$ for δ^2H for MP, SP and ZB samples and $0.05‰$ for $\delta^{18}O$ and $0.13‰$ for δ^2H for NG, LOK and MOD (2014–2018) samples. Samples were re-analyzed in cases where the uncertainty on the three injections lay outside $0.1‰$ for $\delta^{18}O$ and $\leq 0.3‰$ for δ^2H.

Oxygen and hydrogen isotope measurements on the MOD water samples in 2008–2010 were carried out at SILLA (Stable Isotope & Luminescence Laboratory) at the University of Birmingham, UK, while the MOD water samples in 2007–2008 campaign were analyzed by Thermo-Finnigan DELTAplusXP isotope ratio mass spectrometer, fitted with a gas bench and an autosampler at SILab Rijeka (Stable Isotope Laboratory at the Physics Department, School of Medicine, University of Rijeka). Results were normalized using VSMOW-calibrated SILab in-house standards (DZW, RTW, MGS and AAS reported in Mandić et al., [88]) and analytical reproducibility in dual inlet measurements were better than $0.1‰$ for $\delta^{18}O$ and $1‰$ for δ^2H. The same facilities were used for the water analyses on SIR, LUK and VEL samples, with the same measurement precision [39].

Stable isotope analyses of the water samples from GB, DB and DC caves were performed at the Institute for Geological and Geochemical Research (IGGR) in Budapest, Hungary, using a Liquid-Water Isotope Analyser-24d manufactured by Los Gatos Research. The instrument uses off−axis integrated cavity ring down spectroscopy to measure the absolute abundances of $^2H^1H^{16}O$, $^1H^1H^{18}O$ and $^1H^1H^{16}O$ via laser absorption. Measured isotopic ratios were normalized to the homemade laboratory standards, which were calibrated to international standards ($\delta^{18}O = -0.53‰$; $-10.41‰$; $-19.95‰$; $\delta^2H = -9.0‰$; $-74.9‰$; $-147.7‰$ for BWS1, BWS2, BWS3, respectively [89]. The precisions were better than $0.15‰$ and $1.0‰$ for $\delta^{18}O$ and δ^2H, respectively.

Water samples from MED, KRA, VRD, VML and MML caves were measured at the Geological Survey of Israel (GSI). Hydrogen isotope composition was measured by Thermo-Finnigan HighTemperature Conversion Elemental Analyzer (TC/EA) attached to a Delta V Thermo-Finnigan mass spectrometer and oxygen isotope composition was measured by Thermo-Finnigan GasBenchII attached to a Delta Plus mass spectrometer. The precision of these measurements was $\pm 0.1‰$ for $\delta^{18}O$ and $\pm 1.5‰$ for δ^2H.

Modern calcite from MP and SP caves i.e., carbon and oxygen stable isotope composition was analyzed using an Analytical Precision AP2003 continuous-flow isotope-ratio mass spectrometer at the School of Geography, The University of Melbourne, using the procedure described in

Drysdale et al., [90]. Sample results were normalized to the V−PDB scale using an in−house standard of Carrara Marble (NEW1) previously calibrated against the international standards, NBS18 (δ^{13}C = 5.01‰; δ^{18}O = −23.01‰) and NBS19 (δ^{13}C = 1.95‰; δ^{18}O = −2.20‰). Analytical uncertainty was better than 0.1‰ for δ^{18}O and 0.05‰ for δ^{13}C. DB and GB modern calcite samples were analyzed using an automated GasBench II sample preparation device attached to a Thermo−Finnigan DELTAplusXP mass spectrometer at the IGGR in Budapest, Hungary. Precision was better than 0.1‰ for both δ^{18}O and δ^{13}C. The same spectrometer was used for MOD (2008) modern calcite samples at SILab Rijeka and analytical reproducibility for both δ^{18}O and δ^{13}C was better than 0.1‰.

References

1. Brantley, S.L.; Goldhaber, M.B.; Ragnarsdottir, K.V. Crossing Disciplines and Scales to Understand the Critical Zone. *Elements* **2007**, *3*, 307–314. [CrossRef]
2. Regattieri, E.; Zanchetta, G.; Isola, I.; Zanella, E.; Drysdale, R.N.; Hellstrom, J.C.; Zerboni, A.; Dallai, L.; Tema, E.; Lanci, L.; et al. Holocene Critical Zone dynamics in an Alpine catchment inferred from a speleothem multiproxy record: Disentangling climate and human influences. *Sci. Rep.* **2019**, *9*, 1–9. [CrossRef] [PubMed]
3. Fairchild, I.J.; Baker, A. *Speleothem Science: From Process to Past Environments*; Wiley-Blackwell: Chichester, UK, 2012.
4. Hendy, C.H.; Wilson, A.T. Palaeoclimatic Data from Speleothems. *Nature* **1968**, *219*, 48–51. [CrossRef]
5. Hendy, C.H. The Isotopic Geochemistry of Speleothems—I. The Calculation of the Effects of Different Modes of Formation on the Isotopic Composition of Speleothems and Their Applicability as Palaeoclimatic Indicators. *Geochim. Cosmochim. Acta* **1971**, *35*, 801–824. [CrossRef]
6. Henderson, G.M. CLIMATE: Caving In to New Chronologies. *Science* **2006**, *313*, 620–622. [CrossRef]
7. McDermott, F. Palaeo-Climate Reconstruction from Stable Isotope Variations in Speleothems: A Review. *Quat. Sci. Rev.* **2004**, *23*, 901–918. [CrossRef]
8. Badino, G. Models of temperature, entropy production and convective airflow in caves. *Geol. Soc. London Spec. Publ.* **2018**, *466*, 359–379. [CrossRef]
9. Baldini, J.; McDermott, F.; Fairchild, I.J. Spatial Variability in Cave Drip Water Hydrochemistry: Implications for Stalagmite Paleoclimate Records. *Chem. Geol.* **2006**, *235*, 390–404. [CrossRef]
10. Fuller, L.; Baker, A.; Fairchild, I.J.; Spötl, C.; Marca-Bell, A.; Rowe, P.; Dennis, P.F. Isotope Hydrology of Dripwaters in a Scottish Cave and Implications for Stalagmite Palaeoclimate Research. *Hydrol. Earth Syst. Sci.* **2008**, *12*, 1065–1074. [CrossRef]
11. Kim, S.-T.; O'Neil, J.R. Equilibrium and nonequilibrium oxygen isotope effects in synthetic carbonates. *Geochim. Cosmochim. Acta* **1997**, *61*, 3461–3475. [CrossRef]
12. Hartmann, A.; Baker, A. Modelling Karst Vadose Zone Hydrology and Its Relevance for Paleoclimate Reconstruction. *Earth Sci. Rev.* **2017**, *172*, 178–192. [CrossRef]
13. Banak, A.; Mandić, O.; Kovačić, M.; Pavelić, D. Late Pleistocene Climate History of the Baranja Loess Plateau—Evidence from the Zmajevac Loess-Paleosol Section (Northeastern Croatia). *Geol. Croat.* **2012**, *65*, 411–422. [CrossRef]
14. Wacha, L.; Galović, L.; Koloszár, L.; Magyari, Á.; Chikán, G.; Marsi, I. The Chronology of the Šarengrad II Loess-Palaeosol Section (Eastern Croatia). *Geol. Croat.* **2013**, *66*, 191–203. [CrossRef]
15. Wacha, L.; Rolf, C.; Hambach, U.; Frechen, M.; Galović, L.; Duchoslav, M. The Last Glacial Aeolian Record of the Island of Susak (Croatia) as Seen from a High-Resolution Grain-Size and Rock Magnetic Analysis. *Quat. Int.* **2018**, *494*, 211–224. [CrossRef]
16. Durn, G.; Rubinić, V.; Wacha, L.; Patekar, M.; Frechen, M.; Tsukamoto, S.; Tadej, N.; Husnjak, S. Polygenetic Soil Formation on Late Glacial Loess on the Susak Island Reflects Paleo-Environmental Changes in the Northern Adriatic Area. *Quat. Int.* **2018**, *494*, 236–247. [CrossRef]
17. Horvatinčić, N.; Sironić, A.; Barešić, J.; Sondi, I.; Krajcar Bronić, I.; Borković, D. Mineralogical, Organic and Isotopic Composition as Palaeoenvironmental Records in the Lake Sediments of Two Lakes, the Plitvice Lakes, Croatia. *Quat. Int.* **2018**, *494*, 300–313. [CrossRef]
18. Galović, I.; Mihalić, K.C.; Ilijanić, N.; Miko, S.; Hasan, O. Diatom Responses to Holocene Environmental Changes in a Karstic Lake Vrana in Dalmatia (Croatia). *Quat. Int.* **2018**, *494*, 167–179. [CrossRef]

19. Ilijanić, N.; Miko, S.; Hasan, O.; Bakrač, K. Holocene Environmental Record from Lake Sediments in the Bokanjačko Blato Karst Polje (Dalmatia, Croatia). *Quat. Int.* **2018**, *494*, 66–79. [CrossRef]

20. Bakrač, K.; Ilijanić, N.; Miko, S.; Hasan, O. Evidence of Sapropel S1 Formation from Holocene Lacustrine Sequences in Lake Vrana in Dalmatia (Croatia). *Quat. Int.* **2018**, *494*, 5–18. [CrossRef]

21. Felja, I.; Fontana, A.; Furlani, S.; Bajraktarević, Z.; Paradžik, A.; Topalović, E.; Rossato, S.; Ćosović, V.; Juračić, M. Environmental Changes in the Lower Mirna River Valley (Istria, Croatia) during Upper Holocene. *Geol. Croat.* **2015**, *68*. [CrossRef]

22. Kaniewski, D.; Marriner, N.; Morhange, C.; Rius, D.; Carre, M.-B.; Faivre, S.; Van Campo, E. Croatia's Mid-Late Holocene (5200-3200 BP) Coastal Vegetation Shaped by Human Societies. *Quat. Sci. Rev.* **2018**, *200*, 334–350. [CrossRef]

23. Brunović, D.; Miko, S.; Ilijanić, N.; Peh, Z.; Hasan, O.; Kolar, T.; Šparica Miko, M.; Razum, I. Holocene Foraminiferal and Geochemical Records in the Coastal Karst Dolines of Cres Island, Croatia. *Geol. Croat.* **2019**, *72*, 19–42. [CrossRef]

24. Faivre, S.; Bakran-Petricioli, T.; Barešić, J.; Horvatić, D.; Macario, K. Relative Sea-Level Change and Climate Change in the Northeastern Adriatic during the Last 1.5 Ka (Istria, Croatia). *Quat. Sci. Rev.* **2019**, *222*, 1–17. [CrossRef]

25. Surić, M. Speleothem-Based Quaternary Research in Croatian Karst—A Review. *Quat. Int.* **2018**, *490*, 113–122. [CrossRef]

26. Krajcar Bronić, I.; Horvatinčić, N.; Obelić, B. Two Decades of Environmental Isotope Records in Croatia: Reconstruction of the Past and Prediction of Future Levels. *Radiocarbon* **1998**, *40*, 399–416. [CrossRef]

27. Krajcar Bronić, I.; Vreča, P.; Horvatinčić, N.; Barešić, J.; Obelić, B. Distribution of hydrogen, oxygen and carbon isotopes in the atmosphere of Croatia and Slovenia. *Arch. Ind. Hyg. Toxicol.* **2006**, *57*, 23–29.

28. Krajcar Bronić, I.; Barešić, J.; Borković, D.; Sironić, A.; Mikelić, I.L.; Vreča, P. Long-Term Isotope Records of Precipitation in Zagreb, Croatia. *Water* **2020**, *12*, 226. [CrossRef]

29. Horvatinčić, N.; Krajcar Bronić, I.; Obelić, B.; Bistrović, R. Long-time atmospheric tritium record in Croatia. *Acta Geol. Hung.* **1996**, *39*, 81–84.

30. Horvatinčić, N.; Krajcar Bronić, I.; Barešić, J.; Obelić, B.; Vidič, S. Tritium and stable isotope distribution in the atmosphere at the coastal region of Croatia. In *Isotopic Composition of Precipitation in the Mediterranean Basin in Relation to Air Circulation Patterns and Climate*; IAEA-TECDOC-1453; Gourcy, L., Ed.; IAEA: Vienna, Austria, 2005; pp. 37–50.

31. Barešić, J.; Horvatinčić, N.; Krajcar Bronić, I.; Obelić, B.; Vreča, P. Stable Isotope Composition of Daily and Monthly Precipitation in Zagreb. *Isot. Environ. Health Stud.* **2006**, *42*, 239–249. [CrossRef]

32. Vreča, P.; Krajcar Bronić, I.; Horvatinčić, N.; Barešić, J. Isotopic Characteristics of Precipitation in Slovenia and Croatia: Comparison of Continental and Maritime Stations. *J. Hydrol.* **2006**, *330*, 457–469. [CrossRef]

33. Brkić, Ž.; Kuhta, M.; Hunjak, T.; Larva, O. Regional Isotopic Signatures of Groundwater in Croatia. *Water* **2020**, *12*, 1983. [CrossRef]

34. Peel, M.C.; Finlayson, B.L.; McMahon, T.A. Updated World Map of the Köppen-Geiger Climate Classification. *Hydrol. Earth Syst. Sci.* **2007**, *11*, 1633–1644. [CrossRef]

35. Surić, M.; Lončarić, R.; Lončar, N.; Buzjak, N.; Bajo, P.; Drysdale, R.N. Isotopic Characterization of Cave Environments at Varying Altitudes on the Eastern Adriatic Coast (Croatia)—Implications for Future Speleothem-Based Studies. *J. Hydrol.* **2017**, *545*, 367–380. [CrossRef]

36. Czuppon, G.; Bočić, N.; Buzjak, N.; Óvári, M.; Molnár, M. Monitoring in the Barać and Lower Cerovačka Caves (Croatia) as a Basis for the Characterization of the Climatological and Hydrological Processes That Control Speleothem Formation. *Quat. Int.* **2018**, *494*, 52–65. [CrossRef]

37. Rudzka, D.; Mcdermott, F.; Surić, M. A Late Holocene Climate Record in Stalagmites from Modrič Cave (Croatia): Holocene Climate Record from Croatian Stalagmites. *J. Quat. Sci.* **2012**, *27*, 585–596. [CrossRef]

38. Surić, M.; Lončarić, R.; Bočić, N.; Lončar, N.; Buzjak, N. Monitoring of Selected Caves as a Prerequisite for the Speleothem-Based Reconstruction of the Quaternary Environment in Croatia. *Quat. Int.* **2018**, *494*, 263–274. [CrossRef]

39. Paar, D.; Mance, D.; Stroj, A.; Pavić, M. Northern Velebit (Croatia) Karst Hydrological System: Results of a Preliminary ^{2}H and ^{18}O Stable Isotope Study. *Geol. Croat.* **2019**, *72*, 205–213. [CrossRef]

40. Lončar, N. Isotopic Composition of the Speleothems from the Eastern Adriatic Islands Caves as an Indicator of Paleoenvironmental Changes. Ph.D. Thesis, University of Zagreb, Zagreb, Croatia, 2012. (In Croatian)

41. Lončar, N.; Bar-Matthews, M.; Ayalon, A.; Faivre, S.; Surić, M. Early and Mid-Holocene environmental conditions in the Eastern Adriatic recorded in speleothems from Mala špilja Cave and Vela špilja Cave (Mljet Island, Croatia). *Acta Carsol.* **2017**, *46*, 229–249. [CrossRef]

42. Lončar, N.; Bar-Matthews, M.; Ayalon, A.; Faivre, S.; Surić, M. Holocene Climatic Conditions in the Eastern Adriatic Recorded in Stalagmites from Strašna Peć Cave (Croatia). *Quat. Int.* **2019**, *508*, 98–106. [CrossRef]

43. Surić, M.; Roller-Lutz, Z.; Mandić, M.; Krajcar Bronić, I.; Juračić, M. Modern C, O, and H Isotope Composition of Speleothem and Dripwater from Modrič Cave, Eastern Adriatic Coast (Croatia). *Int. J. Speleol.* **2010**, *39*, 91–97. [CrossRef]

44. Bognar, A.; Faivre, S.; Buzjak, N.; Pahernik, M.; Bočić, N. Recent Landform Evolution in the Dinaric and Pannonian Regions of Croatia. In *Recent Landform Evolution*; Lóczy, D., Stankoviansky, M., Kotarba, A., Eds.; Springer: Dordrecht, The Netherlands, 2012; pp. 313–344. [CrossRef]

45. Hartmann, A.; Gleeson, T.; Rosolem, R.; Pianosi, F.; Wada, Y.; Wagener, T. A Large-Scale Simulation Model to Assess Karstic Groundwater Recharge over Europe and the Mediterranean. *Geosci. Model. Dev.* **2015**, *8*, 1729–1746. [CrossRef]

46. Bočić, N. Krš—definicija, svojstva, distribucija. In *Speleologija*, 2nd ed.; Rnjak, G., Ed.; Speleološko društvo Velebit, Hrvatski planinarski savez, Hrvatska gorska služba spašavanja: Zagreb, Croatia, 2017; pp. 557–570.

47. Selak, L. Monitoring okolišnih parametara u turistički uređenoj špilji—primjer Baraćevih špilja kod Rakovice. Master's Thesis, University of Zagreb, Zagreb, Croatia, February 2019.

48. Paar, D.; Ujević, M.; Bakšić, D.; Lacković, D.; Čop, A.; Radolić, V. Physical and Chemical Research in Velebita Pit (Croatia). *Acta Carsol.* **2008**, *37*, 273–278. [CrossRef]

49. Paar, D.; Buzjak, N.; Bakšić, D.; Radolić, V. Physical research in Croatia's deepest cave system Lukina jama-Trojama, Mt.Velebit. In Proceedings of the 16th International Congress of Speleology, Brno, Czech Republic, 21–28 July 2013.

50. Paar, D.; Dubovečak, V. Exploration of deep pits of the Northern Velebit National Park. In *Scientific Report for Northern Velebit National Park*; NP Sjeverni Velebit: Krasno, Croatia, 2014; p. 35.

51. Stroj, A.; Paar, D. Water and Air Dynamics within a Deep Vadose Zone of a Karst Massif: Observations from the Lukina Jama–Trojama Cave System (−1431 m) in Dinaric Karst (Croatia). *Hydrol. Process.* **2019**, *33*, 551–561. [CrossRef]

52. CMHS, 2020 Croatian Meteorological and Hydrological Service.

53. Thornthwaite, C.W. An Approach toward a Rational Classification of Climate. *Geogr. Rev.* **1948**, *38*, 55. [CrossRef]

54. McCabe, G.J.; Markstrom, S.L. *A Monthly Water-Balance Model Driven by a Graphical User Interface*; USGS Open File Report: Reston, VA, USA, 2007.

55. Domínguez-Villar, D. Heat flux. In *Speleothem Science: From Process to Past Environments*; Fairchild, I.J., Baker, A., Eds.; Wiley-Blackwell: Chichester, UK, 2012; pp. 137–145.

56. Gröning, M.; Lutz, H.O.; Roller-Lutz, Z.; Kralik, M.; Gourcy, L.; Pöltenstein, L. A Simple Rain Collector Preventing Water Re-Evaporation Dedicated for $\delta^{18}O$ and δ^2H Analysis of Cumulative Precipitation Samples. *J. Hydrol.* **2012**, *448–449*, 195–200. [CrossRef]

57. Domínguez-Villar, D.; Lojen, S.; Krklec, K.; Kozdon, R.; Edwards, R.L.; Cheng, H. Ion Microprobe $\delta^{18}O$ Analyses to Calibrate Slow Growth Rate Speleothem Records with Regional $\delta^{18}O$ Records of Precipitation. *Earth Planet. Sci. Lett.* **2018**, *482*, 367–376. [CrossRef]

58. Baker, A.; Barnes, W.; Smart, P. Variations in the discharge and organic matter content of stalagmite drip waters in Lower Cave, Bristol. *Hydrol. Process.* **1997**, *11*, 1541–1555. [CrossRef]

59. Collister, C.; Mattey, D. High Resolution Measurement of Water Drip Rates in Caves Using an Acoustic Drip Counter. *AGU Fall Meet.* **2005**, *31*.

60. Scholz, D.; Frisia, S.; Borsato, A.; Spötl, C.; Fohlmeister, J.; Mudelsee, M.; Miorandi, R.; Mangini, A. Holocene Climate Variability in North-Eastern Italy: Potential Influence of the NAO and Solar Activity Recorded by Speleothem Data. *Clim. Past* **2012**, *8*, 1367–1383. [CrossRef]

61. Wassenburg, J.A.; Immenhauser, A.; Richter, D.K.; Niedermayr, A.; Riechelmann, S.; Fietzke, J.; Scholz, D.; Jochum, K.P.; Fohlmeister, J.; Schröder-Ritzrau, A.; et al. Moroccan Speleothem and Tree Ring Records Suggest a Variable Positive State of the North Atlantic Oscillation during the Medieval Warm Period. *Earth Planet. Sci. Lett.* **2013**, *375*, 291–302. [CrossRef]

62. Baker, A.; Hellstrom, J.; Kelly, B.F.J.; Mariethoz, G.; Trouet, V. A Composite Annual-Resolution Stalagmite Record of North Atlantic Climate over the Last Three Millennia. *Sci. Rep.* **2015**, *5*, 10307. [CrossRef]

63. Luetscher, M.; Boch, R.; Sodemann, H.; Spötl, C.; Cheng, H.; Edwards, R.L.; Frisia, S.; Hof, F.; Müller, W. North Atlantic Storm Track Changes during the Last Glacial Maximum Recorded by Alpine Speleothems. *Nat. Commun.* **2015**, *6*, 6344. [CrossRef]

64. Gat, J.R.; Carmi, I. Effect of climate changes on the precipitation patterns and isotopic composition of water in a climate transition zone: Case of the Eastern Mediterranean Sea area. In *The Influence of Climate Change and Climatic Variability on the Hydrologic Regime and Water Resources, Proceedings of the Vancouver Symposium, Vancouver, Canada, August 1987*; IAHS Publication: Vancouver, WA, Canada; pp. 513–523.

65. Gat, J.R.; Klein, B.; Kushnir, Y.; Roether, W.; Wernli, H.; Yam, R.; Shemesh, A. Isotope Composition of Air Moisture over the Mediterranean Sea: An Index of the Air-Sea Interaction Pattern. *Tellus B Chem. Phys. Meteorol.* **2003**, *55*, 953–965. [CrossRef]

66. Dansgaard, W. Stable Isotopes in Precipitation. *Tellus* **1964**, *16*, 436–468. [CrossRef]

67. Celle-Jeanton, H.; Travi, Y.; Blavoux, B. Isotopic Typology of the Precipitation in the Western Mediterranean Region at Three Different Time Scales. *Geophys. Res. Lett.* **2001**, *28*, 1215–1218. [CrossRef]

68. Clark, I.D.; Fritz, P. *Environmental Isotopes in Hydrogeology*, 1st ed.; CRC Press: Boca Raton, FL, USA, 1997. [CrossRef]

69. Yonge, C.J.; Ford, D.C.; Gray, J.; Schwarcz, H.P. Stable Isotope Studies of Cave Seepage Water. *Chem. Geol. Isot. Geosci. Sect.* **1985**, *58*, 97–105. [CrossRef]

70. Bar-Matthews, M.; Ayalon, A.; Matthews, A.; Sass, E.; Halicz, L. Carbon and Oxygen Isotope Study of the Active Water-Carbonate System in a Karstic Mediterranean Cave: Implications for Paleoclimate Research in Semiarid Regions. *Geochim. Cosmochim. Acta* **1996**, *60*, 337–347. [CrossRef]

71. Van Beynen, P.; Febbroriello, P. Seasonal Isotopic Variability of Precipitation and Cave Drip Water at Indian Oven Cave, New York. *Hydrol. Process.* **2006**, *20*, 1793–1803. [CrossRef]

72. Beddows, P.A.; Mandić, M.; Ford, D.C.; Schwarcz, H.P. Oxygen and Hydrogen Isotopic Variations between Adjacent Drips in Three Caves at Increasing Elevation in a Temperate Coastal Rainforest, Vancouver Island, Canada. *Geochim. Cosmochim. Acta* **2016**, *172*, 370–386. [CrossRef]

73. Markowska, M.; Baker, A.; Treble, P.C.; Andersen, M.S.; Hankin, S.; Jex, C.N.; Tadros, C.V.; Roach, R. Unsaturated Zone Hydrology and Cave Drip Discharge Water Response: Implications for Speleothem Paleoclimate Record Variability. *J. Hydrol.* **2015**, *529*, 662–675. [CrossRef]

74. White, W.B. Conceptual models for karstic aquifers. In *Karst Modelling: Special Publication 5*; Palmer, A.N., Palmer, M.V., Sasowsky, I.D., Eds.; Karst Waters Institute Special Publication, The Karst Waters Institute: Charles Town, WV, USA, 1999; pp. 11–16.

75. Ford, D.; Williams, P. *Karst Hydrogeology and Geomorphology*; John Wiley & Sons: Chichester, UK, 2007; p. 578.

76. Smart, P.L.; Friedrich, H. Water movement and storage in the unsaturated zone of a maturely karstified aquifer, Mendip Hills, England. In Proceedings of the Environmental Problems in Karst Terrains and Their Solutions Conference, Bowling Green, KY, USA, 28–30 October 1986; National Water Well Association: Bowling Green, KY, USA, 1987; pp. 57–87.

77. Baker, A.; Hartmann, A.; Duan, W.; Hankin, S.; Comas-Bru, L.; Cuthbert, M.O.; Treble, P.C.; Banner, J.; Genty, D.; Baldini, L.M.; et al. Global Analysis Reveals Climatic Controls on the Oxygen Isotope Composition of Cave Drip Water. *Nat. Commun.* **2019**, *10*, 2984–2987. [CrossRef]

78. Buzjak, N.; Bočić, N.; Paar, D.; Bakšić, D.; Dubovečak, V. Ice Caves in Croatia. *Ice Caves* **2018**, 335–369. [CrossRef]

79. Sharp, Z. *Principles of Stable Isotope Geochemistry*, 1st ed.; Pearson Prentice Hall: Upper Saddle River, NJ, USA, 2007.

80. Rozanski, K.; Araguás-Araguás, L.; Gonfiantini, R. Isotopic patterns in modern global precipitation. *Geoph. Monog.* **1993**, *78*, 1–36.

81. Johnston, V.E.; Borsato, A.; Spötl, C.; Frisia, S.; Miorandi, R. Stable isotopes in caves over altitudinal gradients: Fractionation behaviour and inferences for speleothem sensitivity to climate change. *Clim. Past* **2013**, *9*, 99–118. [CrossRef]

82. Nehme, C.; Verheyden, S.; Nader, F.H.; Adjizian-Gerard, J.; Genty, D.; De Bondt, K.; Minster, B.; Salem, G.; Verstraten, D.; Clayes, P. Cave dripwater isotopic signals related to the altitudinal gradient of Mount-Lebanon: Implication for speleothem studies. *Int. J. Speleol.* **2019**, *48*, 63–74. [CrossRef]

83. Coplen, T.B.; Kendall, C.; Hopple, J. Comparison of stable isotope reference samples. *Nature* **1983**, *302*, 236–238. [CrossRef]

84. Tremaine, D.M.; Froelich, P.N.; Wang, Y. Speleothem calcite farmed in situ: Modern calibration of $\delta^{18}O$ and $\delta^{13}C$ paleoclimate proxies in a continuously-monitored natural cave system. *Geochim. Cosmochim. Acta* **2011**, *75*, 4929–4950. [CrossRef]

85. Daëron, M.; Drysdale, R.N.; Peral, M.; Huyghe, D.; Blamart, D.; Coplen, T.B.; Lartaud, F.; Zanchetta, G. Most Earth-surface calcites precipitate out of isotopic equilibrium. *Nat. Commun.* **2019**, *10*, 429. [CrossRef]

86. Friedman, I.; O'Neil, J.R. Compilation of stable isotope fractionation factors of geochemical interest. In *Data of Geochemistry*, 6th ed.; Fleisher, M., Chap, K.K., Eds.; US Geological Survey Professional Paper: Washington, DC, USA, 1977; Volume 440, pp. 1–12.

87. Coplen, T.B. Calibration of the calcite-water oxygen-isotope geothermometer at Devils Hole, Nevada, a natural laboratory. *Geochim. Cosmochim. Acta* **2007**, *71*, 3948–3957. [CrossRef]

88. Mandić, M.; Bojić, D.; Roller-Lutz, Z.; Lutz, H.O.; Krajcar Bronić, I. Note on the spring region of Gacka River (Croatia). *Isot. Environ. Health S.* **2008**, *44*, 201–208. [CrossRef]

89. Czuppon, G.; Demény, A.; Leél-Őssy, S.; Óvari, M.; Molnár, M.; Stieber, J.; Kármán, K.; Kiss, K.; Haszpra, L. Cave monitoring in Béke and Baradla Caves (NE Hungary): Implications for condition of formation cave carbonates. *Int. J. Speleol.* **2018**, *47*, 13–28. [CrossRef]

90. Drysdale, R.N.; Hellstrom, J.; Zanchetta, G.; Fallick, A.; Goñi, M.F.S.; Couchoud, I.; McDonald, J.; Maas, R.; Lohmann, G.; Isola, I. Evidence for obliquity forcing of glacial termination II. *Science* **2009**, *325*, 1527–1531. [CrossRef]

Article

Using High-Frequency Water Vapor Isotopic Measurements as a Novel Method to Partition Daily Evapotranspiration in an Oak Woodland

Christopher Adkison [1],*, Caitlyn Cooper-Norris [2], Rajit Patankar [3] and Georgianne W. Moore [1]

[1] Department of Ecology and Conservation Biology, Texas A&M University, College Station, TX 77843, USA; gwmoore@tamu.edu

[2] Texas A&M AgriLife Research, Vernon, TX 76384, USA; Caitlyn.cooper@ag.tamu.edu

[3] National Ecological Observatory Network, Denton, TX 76205, USA; rpatankar@battelleecology.org

* Correspondence: ceadkison@tamu.edu

Received: 27 July 2020; Accepted: 19 October 2020; Published: 22 October 2020

Abstract: Partitioning evapotranspiration (ET) into its constituent fluxes (transpiration (T) and evaporation (E)) is important for understanding water use efficiency in forests and other ecosystems. Recent advancements in cavity ringdown spectrometers (CRDS) have made collecting high-resolution water isotope data possible in remote locations, but this technology has rarely been utilized for partitioning ET in forests and other natural systems. To understand how the CRDS can be integrated with more traditional techniques, we combined stable isotope, eddy covariance, and sap flux techniques to partition ET in an oak woodland using continuous water vapor CRDS measurements and monthly soil and twig samples processed using isotope ratio mass spectrometry (IRMS). Furthermore, we wanted to compare the efficacy of δ^2H versus $\delta^{18}O$ within the stable isotope method for partitioning ET. We determined that average daytime vapor pressure deficit and soil moisture could successfully predict the relative isotopic compositions of soil (δ_e) and xylem (δ_t) water, respectively. Contrary to past studies, δ^2H and $\delta^{18}O$ performed similarly, indicating CRDS can increase the utility of $\delta^{18}O$ in stable isotope studies. However, we found a 41–49% overestimation of the contribution of T to ET (f_T) when utilizing the stable isotope technique compared to traditional techniques (reduced to 4–12% when corrected for bias), suggesting there may be a systematic bias to the Craig-Gordon Model in natural systems.

Keywords: water stable isotopes; ecohydrology; evapotranspiration; eddy covariance; forest hydrology; National Ecological Observatory Network (NEON)

1. Introduction

Evapotranspiration (ET) is an important process in the terrestrial hydrological cycle that accounts for evaporation (E) from the soil, open water, and canopy-intercepted water and transpired (T) water from vegetation. ET constitutes a large percentage of the water cycle in most environments, and up to 95% in arid environments, deeming it an important water flux at a variety of spatial scales [1,2]. An improved understanding of ET partitioning and the contribution of T to ET (f_T) is necessary for refined water resource management and quantification of vegetative responses to climate change and alterations to the carbon cycle. For instance, complex dynamics exist between carbon fluxes and precipitation at multiple scales, and the water-use efficiency of plants varies with ecosystem aridity and vegetation cover [3,4]. Partitioning of ET is necessary because T is seen as a desirable aspect of the water cycle that allows vegetation to grow, while water that is evaporated is generally seen as being "lost" from the system.

Stable isotope methods are a rapidly developing field for *ET* partitioning, which has a high potential for advancement beyond longstanding methods, such as water balance [5] and direct *ET* measures compared with direct measures of *T* [6–8]. In a recent review paper [9], 52 studies were discussed that partitioned *ET* using at least two different techniques, including eddy covariance, sap flow, and/or stable isotopes of either δ^2H or δ^{18}O obtained from periodic water samples that were analyzed for isotopic composition using isotope ratio mass spectrometry (IRMS). Of these 52 studies, 30 were in field and row crops, 13 were in orchards or vineyards, and only nine studies were in naturally vegetated areas, such as forests or shrublands. Of the nine studies discussed that took place in natural areas [9], six studies used a combination of eddy covariance and sap flow to partition *ET*, and one study used stable isotopes in combination with a model, but no study used the combination of stable isotopes, eddy covariance, and sap flow altogether. Only one study to date has used this particular combination [8], and this was in an olive (*Olea europaea*) orchard in Morocco. Furthermore, within the stable isotope method, one study in winter wheat (*Triticum aestivum*) found that this method could adequately partition *ET* using δ^2H, but not for δ^{18}O [10]. The reasons for discrepancies between δ^2H and δ^{18}O have been explored [11] and are mainly due to the equilibrium enrichment factor (α^*) of δ^2H being more sensitive to evaporative enrichment from liquid water to water vapor than that of δ^{18}O.

Some studies have found that eddy covariance (EC) measurements performed well when compared to traditional methods for measuring *ET*, such as the catchment water balance and the Bowen ratio energy balance, respectively [5,6], but *T* has also been reported to be underestimated by sap flow, likely due to issues with scaling up *T* from ring-porous tree species (*Quercus* and *Acer*, for example) and accurately capturing the EC tower footprint [12]. Nighttime transpiration has been noted as an important factor in *ET* studies [13], and the traditional heat dissipation equation assumes that flows reach zero every night [14,15]. In systems where vapor pressure deficit (*VPD*) remains high throughout the night, and adequate soil moisture is present, nighttime flows can introduce a significant amount of error in *T* estimation. Furthermore, some studies have reported diurnal and seasonal differences between EC measurements and other techniques, such as stable isotopes [8], while others have found the estimates to be similar [6].

The use of stable isotopes in ecohydrology research has made significant progress since the inception of mass spectrometry in the 1960s and 1970s [16]. Initial experiments in *ET* partitioning-related research specifically date back to the 1990s with Jean-Pierre Brunel being among the first to fully utilize the technique [17–19]. These older experiments manually trapped atmospheric water vapor for 2–4 h at a time, and this same method is still used for collecting water vapor in more recent studies as well [8,10]. Early work from Brunel [18] that has been expanded upon adequately captures isotopic composition at short temporal scales, but the need for higher resolution, more convenient approaches for sampling water vapor isotopes is apparent for studies lasting longer than just a few days, or when many measurements are needed continuously throughout a single day. These challenges brought about the use of cavity ringdown spectrometers (CRDS) in the field for high temporal resolution capabilities.

There is good agreement between results from CRDS and IRMS studies [20]. Furthermore, there have been relatively few studies that have used the stable isotope method specifically for extended periods of time in natural ecosystems, as most studies published have been over a single growing season [21]. To the best of our knowledge, our study is the first report of *ET* partitioning comparisons using sap flux and eddy covariance, along with high-frequency water vapor isotopic methods over multiple seasons at a natural oak woodland site.

In this study within an oak woodland in Texas, USA, our goal was to develop a novel methodology to partition daily *ET* over an entire growing season using combined IRMS and CRDS approaches. By comparison, sap flux measurements were used to estimate daily transpiration and eddy covariance, and Penman–Monteith models were used to estimate *ET*. The main objectives were: To use the IRMS stable isotope method with high-frequency water vapor isotopic measurements from CRDS to (1) partition *ET* in an oak woodland site using stable isotope, sap-flux, and eddy covariance techniques on days where soil and twig sampling provided proxy values for the isotopic compositions of soil (δ_e) and xylem (δ_t) water; (2) to create δ_e and δ_t predictive models for partitioning daily ET on non-sampling days using a combination of monthly field samples, environmental, and micrometeorological data; (3) to compare the efficacy of δ^2H and δ^{18}O in partitioning ET.

2. Materials and Methods

2.1. Site Description

The study site is located within management unit 75 of Lyndon B. Johnson (LBJ) National Grasslands in Wise County, TX, just north of the city of Decatur, TX. It is a core terrestrial site in the National Ecological Observatory Network's (NEON) Southern Plains domain (D11). LBJ is considered part of the Cross Timbers ecoregion of Texas, and wooded areas consist mostly of mixed oaks (Quercus spp.), elms (Ulmus crassifolia and Ulmus alata), eastern red cedar (Juniperus virginiana), and sugarberry (Celtis laevigata) with a mean canopy height of 3.96 m. The LBJ study site is situated at approximately 250 m above sea level, and the EC tower is located precisely at 33.40123° N, −97.57° W. Dominant tree species in the immediate vicinity of the eddy covariance tower are blackjack oak (Quercus marilandica) and post oak (Quercus stellata).

Soils consist of Keeter very fine sandy loams (1–6 percent slopes) and Duffau-Weatherford Complex (3–8 percent slopes) [22]. The mean annual temperature is 18 °C and peak temperatures usually occur in August, while mean annual precipitation (MAP) is 840 mm·year^{-1} with a bimodal distribution (rainy seasons in late spring and fall) [23]. Mean annual potential evapotranspiration (PET) and runoff are 1785 mm·year^{-1} and 8 mm·year^{-1}, respectively [24]. The site has undergone prescribed burning for two consecutive years (2017 and 2018) during spring as a strategy to control the thick understory of greenbrier (Smilax rotundifolia), though the site was not burned at any time during the study period from May 2019 to January 2020. The total period can be divided into the growing/vegetation season (May to the end of October) and the dormancy period (November to January).

2.2. Micrometeorology

At LBJ, NEON operates an EC tower that is fitted with instruments at four different heights in the woodland canopy, with one set of sensors extending above the canopy as well. These instruments include a sonic anemometer, net radiometer, open-path CO_2/H_2O analyzer, as well as tubing that collects vapor at each level and directs it inside a sheltered unit for isotopic analysis on the Picarro™ CRDS.

Actual evapotranspiration (ET_{EC}) values were converted from the given latent heat flux (λE, W·m^{-2}) values to mass units (mm·day^{-1}) using the latent heat of vaporization of water [25] for every 30 min and then summed over 24 h. Days where there were not at least 40/48 data points available were excluded from the analysis.

The flux footprint area was modeled for May of 2019 using the Flux Footprint Prediction Model (FFP) (found at https://geography.swansea.ac.uk/nkljun/ffp/www/) [26]. Input data for the model include wind speed, wind direction, friction velocity, Obukhov length [27], displacement height, measurement height, and standard deviation of lateral velocity fluctuations after rotation.

Due to EC sensor issues at the beginning of the study, the Penman–Monteith reference evapotranspiration (ET_0; mm·day^{-1}) was calculated to provide ET estimates. The FAO-56 method [28] was utilized at a daily scale. If there were incomplete data (<40 data points per day) for any of the input parameters, then that day was excluded from the analysis. A summary of NEON data products used in this study, along with their associated instrumentation, can be found in Table S1 [23].

2.3. Sap Flow Sensors

To directly measure transpiration in the EC tower footprint, eleven oak trees were fitted with thermal dissipation sap flow sensors [14,15] on 11 May 2019 and ran continually until 31 January 2020. All instrumented trees were either Quercus marilandica or Quercus stellata with a minimum, maximum, and average diameter at breast height (DBH) of 21.5, 54.5, and 32.6 cm, respectively (Table S2). All trees had branches above sensor installation height (~1.3 m) and were not shaded by any nearby tree. Sap flow sensors were custom made in the Moore Ecohydrology Lab at Texas A&M University [14,29]. Data was collected every 30 s and later averaged over 30 min intervals on a CR1000 datalogger (CR1000, Campbell Scientific Inc., Logan, Utah).

The sapwood depth was determined by taking tree cores using an increment borer and partially immersing the fresh cores in a safranin-fucsin dye [30–32]. Where the dye clearly moved from the bottom to the top of the core was considered to be actively conducting sapwood. All trees with sensors had a sapwood radius greater than the sensor depth of 2 cm (Table S2) [33]. Sapwood area ranged from 0.022 to 0.149 m^2, with an average of 0.056 m^2 per tree.

2.3.1. Baseline Corrections

It was determined after initial calculations of *VPD* and sap flow processing that nighttime flows were occurring at the site. *VPD* did not reach zero on a nightly basis, and there was not a consistent baseline for ΔT_m which introduces error into the original thermal dissipation equation. To account for this error and adjust ΔT_m for nighttime flows, a "baseline smoothing" process was utilized where daily minimum *VPD* was plotted with daily precipitation (Figure 1A) and days that had <0.1 kPa *VPD* and zero precipitation were considered days where calculated ΔT_m was accurate. The ΔT_m for remaining days were linearly gap-filled to create a smooth baseline. These new ΔT_m were then used to re-process the sap flow data.

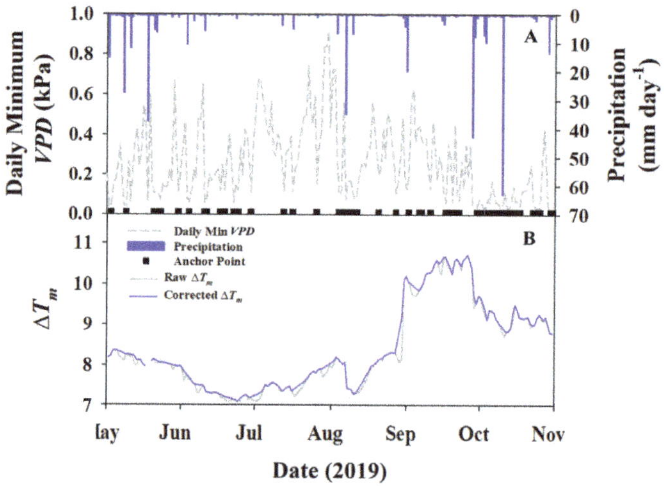

Figure 1. (**A**) Daily minimum vapor pressure deficit (dashed lines) and precipitation (blue bars) with days having a true ΔT_m marked (black squares); (**B**) the raw measured average ΔT_m and corrected ΔT_m portraying baseline smoothing.

2.3.2. Transpiration Scaling

Scaling up sap flow measurements to the footprint of the EC tower was done using surveys from trees in 29 individual 400 m^2 vegetation plots [23]. Within the 29 plots, there were 915 live individuals total for which the stem diameter and tree height were recorded. Using DBH, the basal area was then calculated for each individual tree, and an allometric equation was applied to calculate the sapwood area for each individual based on the basal area (y = 0.6075x + 0.0004, R^2 = 0.92). The total sapwood area per plot was divided by the total plot area to get sapwood area per unit ground area (m^2 m^{-2}) for each plot, and a threshold value of 0.2 m^2 of sapwood area was used to differentiate between woodland and grassland plots (Figure S1). The average sapwood area per unit ground area for all of the woodland plots ($\bar{x}$ = 0.0012 m^2 m^{-2} ± 0.0006, n = 19) was used to calculate the total sapwood area in the EC tower airshed. Transpiration (mm·day^{-1}) was then calculated by multiplying the average sap-flux density of all sap flow trees (kg m^{-2} day^{-1}) by the average sapwood area per unit ground area. This bottom-up approach has been utilized in other studies [5,8] and validated by a review on methods for scaling up T measurements from tree to stand level [34]. It should be noted that this T estimate was based on oak allometry and sap flux, but there were other species present in the plots that may have different sapwood area to basal area ratios. Note that no understory species were included in this T estimate, including any woody vegetation with DBH < 0.5 cm.

2.4. Stable Isotopes

2.4.1. Method Framework

Given $ET = E + T$, the respective contributions of E and T to ET can be calculated using the isotopic mass balance approach, assuming a two-source model

$$\delta_{ET}ET = \delta_E E + \delta_T T \tag{1}$$

where δ_{ET}, δ_E, and δ_T are equal to the isotopic composition of evapotranspiration, soil evaporation, and plant transpiration, respectively. The ratio of T to ET can then be calculated as

$$f_T = \frac{T}{ET} = \frac{\delta_{ET} - \delta_E}{\delta_T - \delta_E} \tag{2}$$

To use this method, the isotopic composition of water in three "end members" (i.e., soil, vegetation, and atmosphere) are needed, and the resulting ratios of each can be applied to total ET to get values of the individual components [35,36]. Differences in the isotopic composition of water in vegetation and in soil occur from fractionation during the evaporation process. The soil evaporating front changes with soil texture, soil moisture, and other environmental conditions, but is generally at the surface when the soil is saturated, and between 0.2–0.3 m in depth when the soil is not saturated [37–39]. Water is not fractionated during the transpiration process under steady-state conditions, so the isotopic composition of transpired water can be assumed equal to that of the xylem water [18,40,41]. Steady-state conditions are usually met around midday in field conditions [42]. Figure 2 shows the stepwise process for partitioning ET using the stable isotope method.

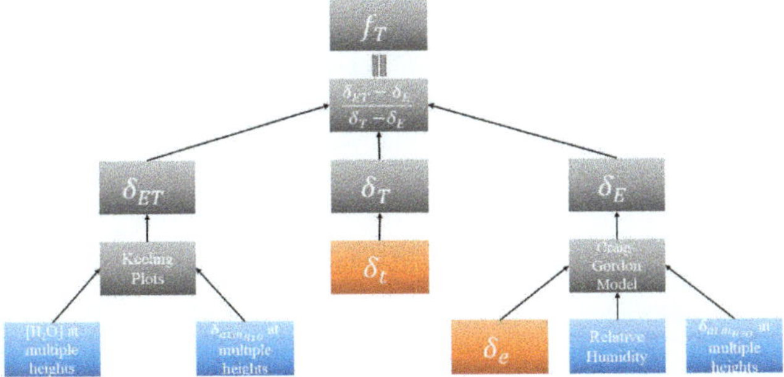

Figure 2. Conceptual diagram for obtaining f_T using the stable isotope method. All δ values represent either δ^2H or δ^{18}O. Blue boxes represent 30 min National Ecological Observatory Network's (NEON) data, orange boxes represent monthly samples. δ_e = isotopic composition of soil water at evaporating front, obtained from soil samples and δ_t = isotopic composition of xylem water, obtained from twig samples collected under steady-state conditions.

2.4.2. δ_{ET}, δ_E, and δ_T

The Keeling plot approach was used for obtaining δ_{ET} [35]. Measurements of the concentration and isotopic composition of water vapor at varying heights in the woodland canopy are sufficient to fit the linear model used in the Keeling plot approach [43–45]. Data for this approach were available from NEON [23]. Water vapor was collected by tubing at five different heights on a 30 min time step in the woodland canopy and circulated into the instrument hut where it was then analyzed for $\delta^{18}O$ and δ^2H via cavity ringdown spectroscopy (Picarro L2130-i, Picarro Inc., Santa Clara, CA, USA) in near-real-time. Water vapor concentrations at the same heights were included in the bundled EC data products. Only values from 10:30–14:00 h were used in the analysis.

Values of δ_E were attained from a combination of field samples and the Craig-Gordon Model (CGM) [46]

$$\delta_E = \frac{\delta_e/\alpha^* - h\delta_v - \varepsilon_{eq} - (1-h)\varepsilon_k}{(1-h) + (1-h)\varepsilon_k/1000} \tag{3}$$

where δ_e is the isotopic composition of liquid water at the soil evaporating front, α^* is the equilibrium fractionation factor (1.0098 and 1.084 at 20 °C for ^{18}O and ^2H, respectively [47]), h is relative humidity, δ_v is the isotopic composition of the background atmospheric water vapor, α_k is the isotopic fractionation factor (0.9755 and 0.9723 for ^2H and ^{18}O, respectively [48,49]), ε_{eq} is equal to $1000(1 - 1/\alpha^*)$, and ε_k is equal to $1000 (\alpha k - 1)$.

Values for δ_T were attained directly through a sampling of twig water under steady-state assumptions, as has been done in other studies [35]. Under this scenario, it is assumed that the isotopic composition of the source water and transpired water are the same ($\delta_X = \delta_T$).

2.5. Field Sampling and Sample Analysis

2.5.1. Soil and Twig Sampling

To obtain isotopic values for soil water at the evaporating front (δ_e) and xylem water (δ_t), monthly soil and twig samples were taken. There was a total of nine sampling events during the study period that were carried out approximately every four weeks (Day of Year [DOY] 143, 170, 204, 234, 266, 294, 329, 351 in 2019 and DOY 24 in 2020). Soil samples (n = 4 per sampling day) were collected at midday (between 10:30–14:00 h) every month during the period of the study (May 2019–January 2020) to capture seasonal variability in δ_E. Samples were collected at 0.2 to 0.3 m depth when the soil was unsaturated, and taken from the surface when the soil was saturated. Soil saturation was determined at the site on sampling days before cores were taken. Samples were taken using a hand auger and immediately transferred into 12 mL glass vials and sealed with a cap and parafilm. To prevent evaporation, vials were transported upside down at ~34 °C to a freezer within 8 h. Samples were kept frozen until being processed at the Stable Isotopes for Biosphere Science (SIBS) Laboratory at Texas A&M University.

Twig samples (n = 4 per sampling day) were collected on the same days as soil samples during the same time window to satisfy steady-state conditions for xylem water sampling and to capture seasonal variability in δ_T. Samples were taken using a pole pruner from mature branches on trees containing sap flow sensors and were approximately three to five centimeters in length and less than 19 mm in diameter. The bark was then stripped from the twigs, and the twigs were immediately transferred into 12 mL glass vials and stored in the same manner as the soil samples.

2.5.2. Cryogenic Water Extractions and Analysis

All samples underwent water extractions using a cryogenic (liquid nitrogen) vacuum distillation system. Soil samples were extracted for approximately 75 min and twigs for approximately 55 min. The extracted water was then transferred to scintillation vials via pipet and stored at 34 °C until analysis.

The hydrogen and oxygen isotope analyses were performed using a Thermo Scientific High Temperature Conversion/Elemental Analyzer (TC/EA) coupled to a Conflo IV and a Thermo Scientific Delta V Advantage IRMS at the Stable Isotopes for Biosphere Science Laboratory, Texas A&M University. One micro liter (1 µL) of the sample was injected into TC/EA using a PAL autosampler. The injected sample was converted to H_2 and CO gas by pyrolysis reaction through a glassy carbon tube filled with glassy carbon chips and heated at 1370 °C. The H_2 and CO gas were separated by a 2 m packed gas chromatograph and were analyzed for the hydrogen and oxygen isotope ratios, respectively, in the Delta V Advantage IRMS. One sample was injected three times, and reported values are an average of these triplicate injections ± 0.01‰.

Calibration curves were derived using in-house water standards: SIBS-wA (δ^2H = −390.8 ‰, $\delta^{18}O$ = −50.10 ‰) and SIBS-wP (δ^2H = −34.1 ‰, $\delta^{18}O$ = −4.60 ‰). Quality control was performed using an in-house water standard, SIBS-wU (δ^2H = −120.3 ‰, $\delta^{18}O$ = −15.91 ‰). These in-house standards were calibrated using IAEA standards (VSMOW2, SLAP, and GISP) in 2009 and 2014.

2.6. f_T Analysis for Sampling Days

To determine how f_T compares between the stable isotope method and the residual of ET (ET_{EC} or ET_0) and T, only data from days when the soil and twig samples were collected (to provide surrogate values for the Keeling plots and CGM) were analyzed initially for the growing season (May–October 2019). Four methods were used to calculate f_T: (1) T/ET_{EC}, (2) T/ET_0, (3) δ^2H-based stable isotope method, and (4) $\delta^{18}O$-based stable isotope method. This approach was used to provide insight as to how various partitioning methods compare to each other, how δ^2H and $\delta^{18}O$ compare within the stable isotope method, as well as how the various methods might change over time and with changes in seasonality. Samples were taken for November 2019 to January 2020 as well; however, it was assumed that f_T would be zero during this time, due to the trees having entered dormancy and the leaves abscising.

3. Predictive Model Development

An advantage of our study relative to others like it is the high data resolution and data availability across spatial and temporal scales that are collected by NEON. Due to the high-resolution data that is available throughout the growing season, two models were developed for predicting δ_e and δ_t for all days in the study period based on data from the sampling days. As obtaining soil and twig samples on a daily basis is cumbersome, expensive, and logistically infeasible, this inherently limits the utility of the stable isotope method by limiting the temporal resolution of surrogate values used in the CGM and Keeling plots. However, this barrier may be removed by predicting the isotopic composition of δ_e and δ_t (both δ^2H and $\delta^{18}O$) based on other environmental variables that are easier to measure and do not require on-site presence.

Single and multiple linear regression models were tested using environmental data provided from NEON (air temperature, *VPD*, soil moisture, precipitation, soil heat flux, net radiation, water vapor concentration, water vapor isotopic composition, wind speed, and relative humidity) to predict the measured isotopic composition of δ_e and δ_t for sampling days during the growing season. Values from the δ_t predictive models were used directly for δ_T in the isotope method for partitioning ET, and values from the δ_e predictive models were used in the CGM to get δ_E. f_T was then determined every day from 11 May to 31 October for both δ^2H and $\delta^{18}O$.

4. Results

4.1. Environmental Data

The majority of the growing season was characterized by high air temperature (>20 °C), stable wind speed, high daytime *VPD* (>0.3 kPa), and low soil moisture (<0.20 cm^3 cm^{-3}) following storms in May and June (Figure 3). Soil moisture fell consistently between July and September, and this same period experienced the highest daily air temperatures and the least amount of precipitation during the study period. Daytime *VPD* also reached its maxima in August, indicating peak summer conditions.

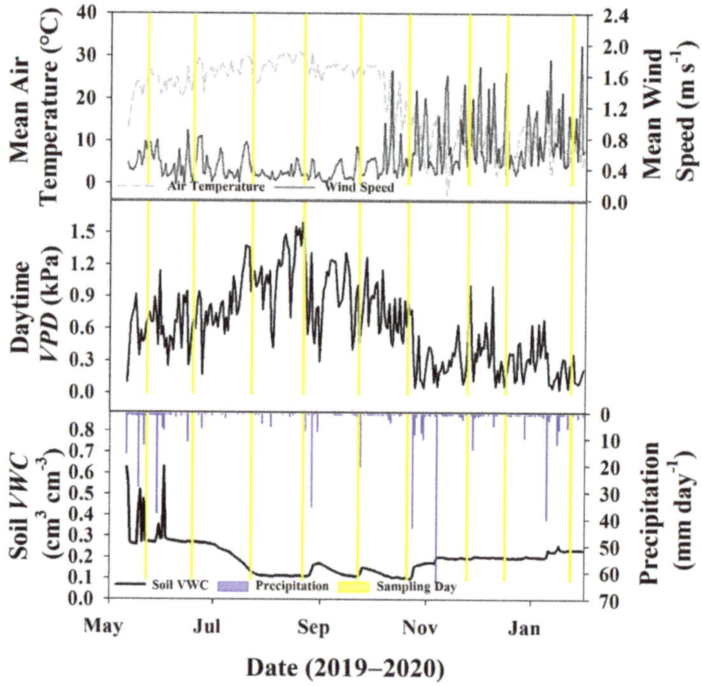

Figure 3. (**A**) Thirty-minute average air temperature and wind speed recorded from the eddy covariance (EC) tower during the study period; (**B**) average daytime (6:00–18:00 h) vapor pressure deficit (*VPD*); (**C**) and soil volumetric water content (*VWC*) at 0.16 m depth at soil plot 1 along with daily precipitation totals. Days where there was an isotope sampling event (soil and twigs) are highlighted in yellow.

However, with the onset of the dormant season at the end of October and the beginning of November, the air temperature dropped markedly, and daily average wind speed showed higher fluctuations without the tree canopy acting as a buffer around the EC tower (Figure 3A). Due to changes with wind speed and air temperature dynamics, daytime *VPD* also began to drop, relative to growing season conditions (Figure 3B). The beginning of the dormant season was also characterized by a few large rain events that elevated soil moisture levels relative to the summer, and smaller, more frequent rain events maintained these higher levels throughout the end of the study (Figure 3C).

Water vapor isotopes also had distinct characteristics during the growing and dormant seasons. While trees were still transpiring, temperatures were still warm, and wind speeds were stable, average water vapor δ^2H and δ^{18}O was also relatively stable (Figure 4). As trees began to senesce in October and November, resulting in less stomatal conductance, the boundary layer conditions changed, and this is reflected in the higher degree of fluctuations of water vapor isotopes in the dormant season relative to the growing season.

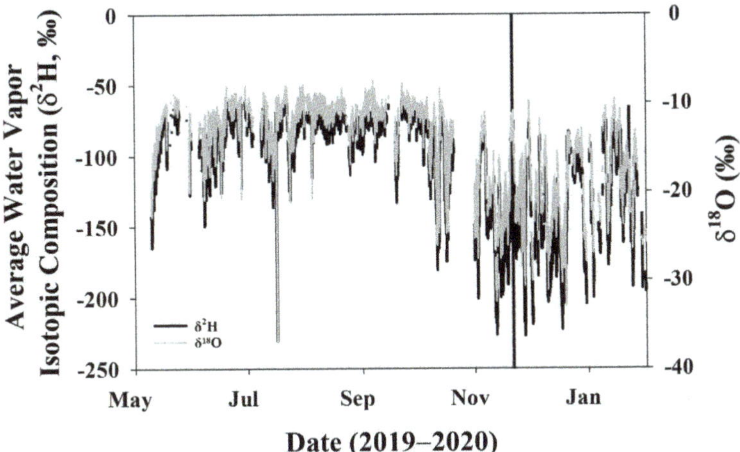

Figure 4. Average isotopic composition of water vapor for the study period. The vertical black line indicates DOY 325 when the trees entered dormancy and leaves abscised.

4.2. Water Fluxes

Due to technical issues, actual evapotranspiration (ET_{EC}) was unreliable for the beginning of the study period (Figure 5). However, there was a moderate agreement for daily evapotranspiration estimated by Penman–Monteith (ET_0) and the eddy covariance method for the remainder of the growing season (2.25 ± 0.89 mm·day^{-1} and 2.60 ± 0.92 mm·day^{-1}, respectively). The eddy covariance method provided sufficient data 40% of the time from 11 May 2019 to 31 January 2020 (107/265 days in the study period), and ET_0 provided sufficient data 88% of the time (233/265 days).

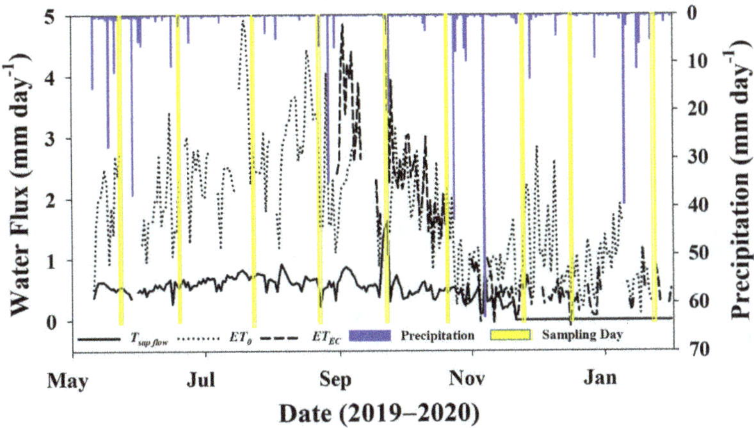

Figure 5. Daily transpiration ($T_{\text{sap flow}}$), reference evapotranspiration (ET_0), actual evapotranspiration (ET_{EC}), and precipitation for the study period. Days where there was an isotope sampling event (soil and twigs) have been highlighted in yellow.

Total precipitation during the measurement period prior to dormancy (11 May to 31 October 2019) was 315.77 mm, or about 38% of MAP, which was 36% below normal for that time of year based on data from the last 50 years [50]. Total T, ET_0, and ET_{EC} for this same time period were 102.2 mm

(12% of MAP), 391.5 mm (47% of MAP), and 452.4 mm (54% of MAP), respectively, for that partial growing season. Average transpiration for the study period was 0.59 ± 0.18 mm·day^{-1}, and trees were officially deemed dormant on DOY 325 (21 November 2019) based on sap flow data, just before the November sampling event. The average ET_0 and ET_{EC} for the dormant season (November–January) was 0.94 ± 0.55 and 0.44 ± 0.28 mm·day^{-1}.

4.3. Sampling Days

4.3.1. Global Meteoric Water Line and Keeling Plot

The isotopic composition ($\delta^{18}O$ and δ^2H) of atmospheric and precipitation samples fell closely along the global meteoric water line (GMWL), as was expected (Figure 6). The soil samples fell to the right of the GMWL, indicating preferential evaporation of lighter isotopologues of water and soil enrichment with the heavier isotopologues. These results indicate the evaporative fractionation of soil water and an appropriate degree of separation between the signals of twig and soil water isotope ratios.

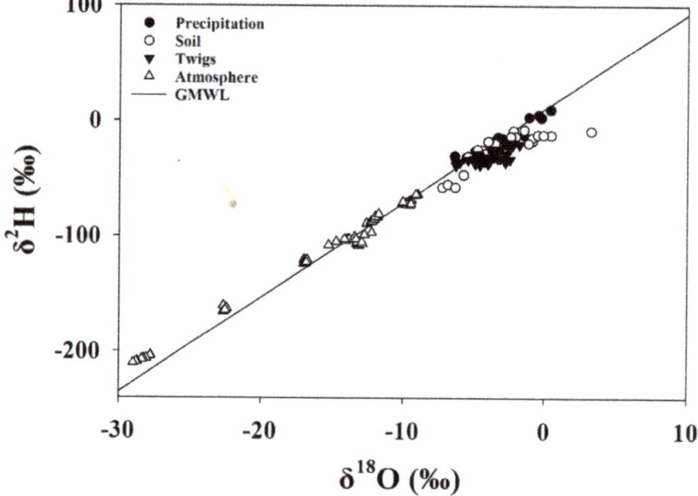

Figure 6. δ^2H versus $\delta^{18}O$ for soil, precipitation, twig, and atmospheric water samples as related to the global meteoric water line (GMWL) for the entire study period.

The average of the inverse water concentration between 10:30–14:00 h was used against the isotope ratio (both $\delta^{18}O$ and δ^2H) during the same time period to construct turbulent mixing relationships, or Keeling plots. In examining Keeling Plot relationships, we found that the inverse water vapor concentration was closely related to isotopic ratios during all sampling events ($p < 0.05$) except July (δ^2H and $\delta^{18}O$) and October ($\delta^{18}O$) with R^2 ranging up to 0.86 (Table 1). As 1/[H$_2$O] increased, heavier isotopes tended to decrease. The monthly predictions of δ_{ET} from Keeling plots varied widely throughout the study period for both H and O.

Table 1. Descriptive statistics for linear regression (Keeling Plot) analyses for sampling events during the growing season. Sample size (*N*) is the total of water vapor data from 10:30 to 14:00 at five different heights in the forest canopy for each sampling day. DOY, Day of Year.

Sampling DOY 2019	Stable Isotope	Keeling Plot				
		N	Slope	δ_{ET}	R^2	*p*-Value
143	$\delta^{18}O$	21	−236.37	−1.07	0.71	<0.001
	δ^2H	21	−1853.90	−3.37	0.77	<0.001
170	$\delta^{18}O$	29	−97.11	−7.86	0.61	<0.001
	δ^2H	29	−864.74	−47.32	0.86	<0.001
204	$\delta^{18}O$	29	109.86	−19.60	0.09	0.11
	δ^2H	29	184.18	−112.54	0.02	0.52
234	$\delta^{18}O$	29	103.95	−13.95	0.13	0.05
	δ^2H	29	−682.05	−44.91	0.57	<0.001
266	$\delta^{18}O$	16	−368.38	3.03	0.44	0.005
	δ^2H	16	−1084.78	−28.37	0.29	0.03
294	$\delta^{18}O$	28	7.11	−13.57	0.03	0.4
	δ^2H	28	−331.99	−76.86	0.72	<0.001

4.3.2. f_T for Soil and Twig Sampling Days

The average f_T calculated from the stable isotope method with data from only sampling days was 0.85 ± 0.27 for $\delta^{18}O$ and 0.7 ± 0.55 for δ^2H. On two of the sampling days where the Keeling Plots had significant slopes (May and September) f_T exceeded 100% of total *ET* for both δ^2H and $\delta^{18}O$. Additionally, the δ^2H July sampling event had a $f_T < 0$, though the Keeling Plot for this event was not significant. The average f_T calculated from ET_0 was substantially lower (0.25 ± 0.1) and never went below 0 or above 1 (Figure 7).

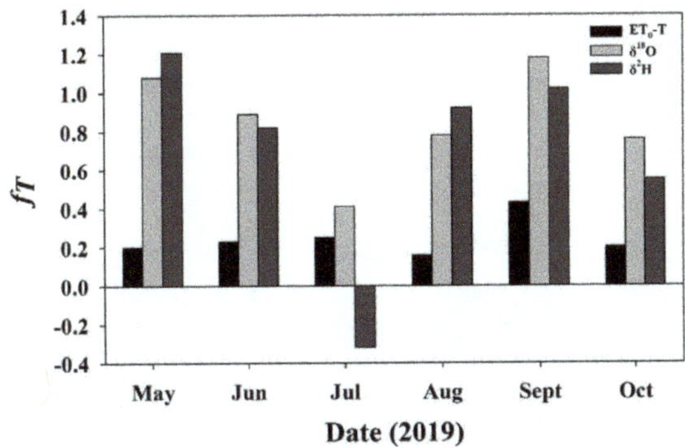

Figure 7. The fractional contribution of transpiration to total evapotranspiration on sampling events during the 2019 growing season calculated using the stable isotope method for both δ^2H (dark grey bars) and $\delta^{18}O$ (light grey bars), and using the residual of T/ET_0 from scaled sap flow measurements (black bars).

4.4. Predictive Models for Growing Season and Dormant Season Sampling Days

It was determined that average daytime *VPD* (between 6:00–18:00 h) and average soil moisture at 0.16 m depth were the single best predictors of δ_e and δ_t, respectively (Figure 8). Soil moisture at other depths was tested; however, 0.16 m performed the best and was empirically logical as it is the depth in-between saturated (surface) and unsaturated (0.2 to 0.3 m) sample depths. Equations from Figure 8 were then used to estimate daily values of δ_e and δ_t for use in the stable isotope method to partition *ET*. All growing season models were significant at the 95% confidence level, with the exception of the δ_e δ^2H model. A total of 6 data points were used in the δ_t models, but the July sampling event was excluded from the δ_e models as its Keeling Plot had an insignificant slope (Table 1), and its isotope ratio values were outliers compared to the other five months.

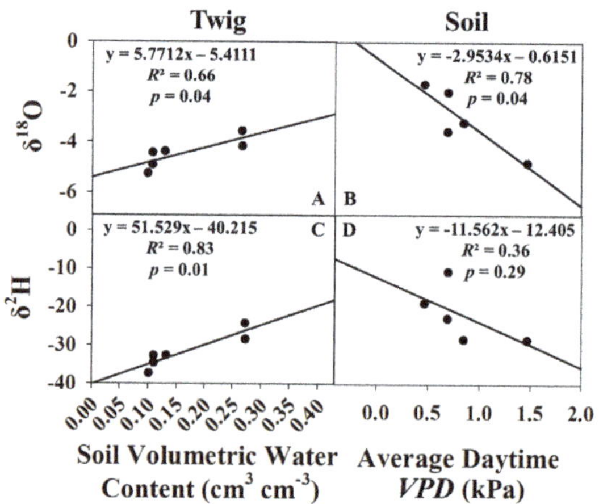

Figure 8. Linear regressions of δ^{18}O vs. average soil moisture (**A**) and average daytime *VPD* (**B**), δ^2H vs. average soil moisture (**C**), and average daytime *VPD* (**D**) for the growing season only.

For the dormant season, the same approach was utilized for δ_e and δ_t modeling during the growing season. The additional data from November 2019–January 2020 was added to the four models, and new regressions were conducted. The δ_t models were mostly unaffected and continued to be significant for both δ^2H and δ^{18}O (Figure 9). However, for the δ_e models, the slope of the regression line switched directions, the R^2 values dropped markedly, and the *p*-value for the previously significant δ_e δ^{18}O model jumped to 0.75. Due to these reasons, δ_e and δ_t were not modeled for the dormant season.

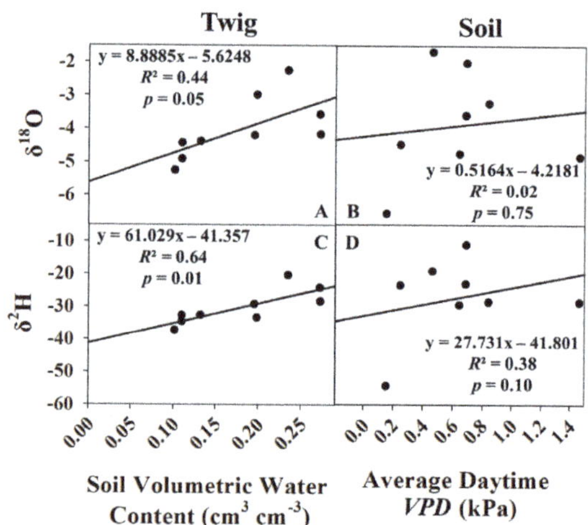

Figure 9. Linear regressions of δ^{18}O vs. average soil moisture (**A**) and average daytime *VPD* (**B**), δ^2H vs. average soil moisture (**C**), and average daytime *VPD* (**D**) for the growing season and dormant season combined.

4.5. Entire Growing Season

4.5.1. Keeling Plot

To predict δ_{ET} for the entire growing season, atmospheric isotope data and water vapor concentration data for every day from 11 May 2019 to 31 October 2019 were collected and filtered for data only during 10:30–14:00 for each of the five heights in EC tower. This data was then input into Rstudio and underwent linear regression (Keeling plots/turbulent mixing relationships) for every day there were sufficient data, for both $\delta^{18}O$ and δ^2H. Relevant parameters were extracted for each day, including sample size (N), slope, intercept (δ_{ET}), R^2, and a p-value, which are reported in Table S3. Of the 164 days in the growing season, there were sufficient EC tower data to construct Keeling plots for 131 days (80%). Of these 131 days with complete data, there were significant slopes for 112 days (86%) using δ^2H, and 89 days (68%) using $\delta^{18}O$. Keeling plots were not constructed for the dormant season, due to issues with δ_e modeling and unstable boundary layer conditions.

4.5.2. Craig-Gordon Model

Of the 164 days in the growing season, the CGM successfully ran for 148 (90%) of days. In this case, $\delta^{18}O$ and δ^2H performed equivalently in terms of producing values for δ_E, and the limiting factor for running the CGM was obtaining complete water vapor isotope data from the EC tower and CRDS. Sixteen days (10%) were missing or had incomplete water vapor isotope data. Relative humidity data were complete for the entire season, as well as soil moisture used to predict δ_e.

4.5.3. Daily f_T for Entire Study Period

Days with insignificant Keeling plots (p-value > 0.05) were excluded for f_T using the stable isotope method, along with days having an $f_T < 0$ or >1.0 for all methods. Day 131 was excluded, due to site setup and disturbance. For the first approach (T/ET_{EC}), the average f_T was 0.36 ± 0.31, and there was a total of 70 days (43%) with valid and complete data. For the second approach (T/ET_0), the average f_T was 0.28 ± 0.11, and there was a total of 140 days (86%) with valid and complete data. Both δ^2H and $\delta^{18}O$ had an average f_T of 0.77 ± 0.20, though there were only valid data for 79 and 56 of the 164 days (48% and 34%), respectively. The results from raw calculated f_T are reported in Figure 10.

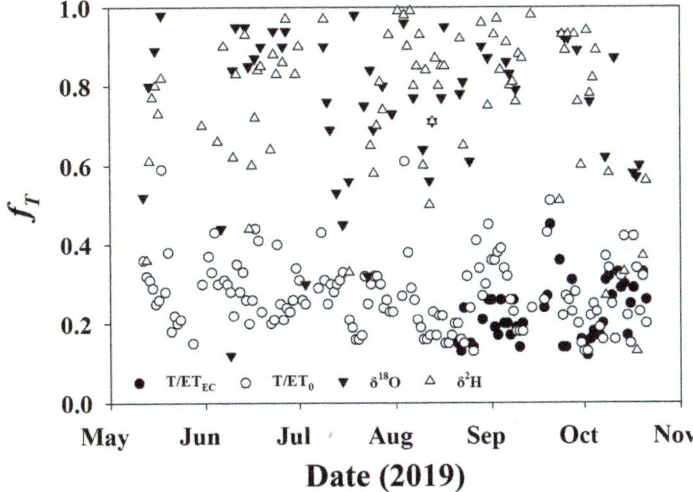

Figure 10. Fractional transpiration computed using T/ET_{EC} (closed circles), T/ET_0 (open circles) δ^2H (open triangles), and $\delta^{18}O$ (closed triangles).

On average, f_T calculated from T/ET_{EC} or T/ET_0 were 41% and 49% lower than those calculated from the stable isotope method. Minimum f_T calculated from the stable isotope method was nearly always greater than the maximum value calculated by traditional methods (T/ET_{EC}, T/ET_0). To remove this potential systematic bias from the estimates, f_T from the stable isotope method was bias-corrected to ET_0 (Figure 11 left) and ET_{EC} (Figure 11 right) using the average difference as the trends (direction and slope) were similar, despite the absolute magnitude being different. The average ET_0 bias-corrected stable isotope f_T was 0.40 ± 0.15, and the ET_{EC} bias-corrected f_T was 0.32 ± 0.15, which are only 4–12% greater than traditional f_T, as opposed to 41–49% before normalization.

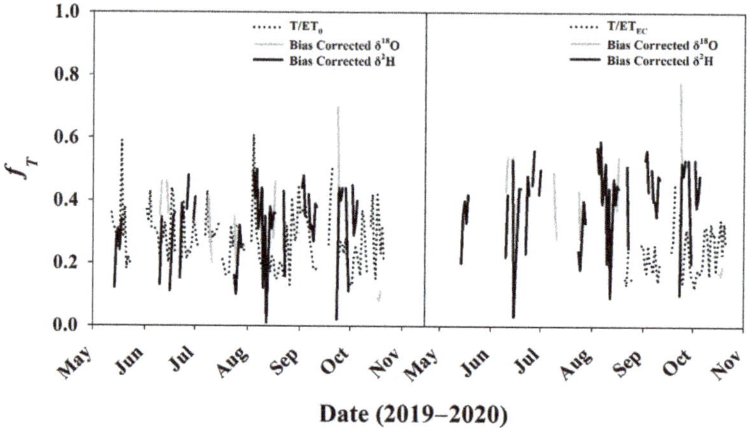

Figure 11. Stable isotope f_T bias-corrected to (**left**) ET_0; (**right**) ET_{EC}. Data for ET_{EC} before DOY 241 were invalid and excluded from the analysis.

5. Discussion

This study demonstrated the utility of using a combination of stable isotopes, sap flux, and eddy covariance techniques to partition ET in an oak woodland and introduces a novel methodology using CRDS water vapor isotopes. The stable isotope technique has benefited from improved technological advancements with cavity ringdown spectroscopy and high-temporal resolution vapor collection systems that work in unison with eddy covariance and sap flux systems. However, the stable isotope method for partitioning ET did not compare favorably with more traditional techniques, such as using the difference of total ET from eddy covariance or ET_0 along with T estimated from sap flow measurements, on any sampling day. There was a 41–49% overestimation of f_T in this system when utilizing the stable isotope technique compared to ET_{EC} or ET_0 relative to sap-flux-based T. When using the average difference to normalize f_T for either $\delta^{18}O$ or δ^2H the overestimation was reduced to 4–12%, which is within the range found from other studies [10]. This suggests that there may be a systematic bias to the CGM, which leads to the overestimation of f_T in natural systems, and this bias should be investigated thoroughly in the future.

When comparing $\delta^{18}O$ and δ^2H within the stable isotope method, there was much agreement between the two, which contrasts with results and conclusions from other studies [10,21,32]. Typically, δ^2H is utilized over $\delta^{18}O$, due to its higher degree of sensitivity. However, the high temporal-resolution data provided by the CRDS was able to provide the necessary data resolution for $\delta^{18}O$ to have a similar performance to δ^2H. Both $\delta^{18}O$ and δ^2H were able to satisfy requirements for the CGM 90% of the time, and $\delta^{18}O$ only slightly underperformed in the Keeling Plot approach when compared to δ^2H (68% and 86% of the study period). This may be in part to the more sensitive fractionation coefficient for δ^2H relative to $\delta^{18}O$, which increases the potential for significant results, or there may have been evaporation in soil and twig sample vials during transit from the field to the lab (for IRMS analysis), which could have led to the error. However, these results complement those from the GMWL, indicating that the stable isotope approach can successfully be utilized within the framework provided in this study. The limiting factors in constructing Keeling plots at the daily scale using the CRDS and the EC tower are having a robust data collection system and a field crew that can maintain and repair the instruments in the case of malfunction or failure. While δ^2H slightly outperformed δ^{18}O, which was expected, δ^{18}O performed more strongly than previous studies have suggested. This indicates that the high temporal-resolution data collected by the CRDS is truly necessary to make this approach viable at the daily scale for both stable isotopes of water.

A severe limiting factor to the utility of the stable isotope approach for partitioning *ET*, however, is the ability to obtain soil and xylem water samples from the field at fine enough resolutions to examine multi-day or seasonal trends in δ_e and δ_t. For this study, field samples were only feasibly obtained monthly, which limited the data available for partitioning initially to one day per month. However, due to partnership with NEON, a diverse suite of data was available at a much higher resolution, and soil and xylem water were successfully modeled using 30 min *VPD* and soil moisture data, respectively. This enabled us to partition *ET* from what was initially only six sampling events, to 79 and 56 days for δ^2H and δ^{18}O, respectively. While this was still only 56% and 48% of all days in the growing season, the stable isotope method was markedly improved by utilizing a combination of modeling and field sampling events. Our model also coincides with results from recent work that established tree water status, driven by soil water potential and atmospheric conditions, was the main reason for source water partitioning and isotopic fractionation during a 7-week lysimeter experiment in Switzerland [51].

While 0.16 m soil depth worked best during the growing season for this study, other depths may work better under different precipitation regimes. For instance, if the soil constantly remained unsaturated during the growing season, then 0.2 to 0.3 m may be more appropriate to model δ_e dynamics. Similarly, under flooded conditions, it would be more appropriate to model δ_e from soil moisture at the surface. While source water for trees was not independently identified during this study, that may be advantageous in the future to verify the assumption that $\delta_X = \delta_T$. Furthermore, modeled δ_t in this study was based solely on xylem water from *Quercus* spp. twig samples. However, these values may change if other species in the tower footprint were sampled instead. A more accurate estimation of both T and δ_t may be achieved by measuring sap flow in additional species and sampling those species for xylem water simultaneously. Steady-state assumptions were justified in this study, due to sampling in the mid-day and afternoon conditions, but there is an increasing body of evidence that in the early morning and evening, steady-state conditions are no longer present. This should be taken into consideration for future work; however, some studies have also shown that f_T calculated both ways is similar at the end of the day [36].

When dormant season data was added to the δ_e and δ_t models, δ_t remained significant as the water inside of the trees were not being utilized during dormancy; however, the δ_e models became unreliable. This suggests that evaporation dynamics during the dormant season are different than during the growing season, and that these dynamics were not successfully captured in the CGM. Furthermore, atmospheric water vapor isotopes at the onset of and during the dormant season are considerably more variable than during the growing season. This is likely due to the physical barrier that leaves give at the top of the canopy, as well as the regulating properties of active transpiration on the boundary layer micrometeorology [52].

It should be noted that forests and woodlands inherently have more variability and heterogeneity when compared to engineered or agricultural systems like croplands, and this may result, in part, account for variance observed in the boundary layer. However, the combination of insignificant δ_e models and sporadic water vapor concentrations just before and after trees lose their leaves suggests that the CGM cannot be utilized during the dormant season in such systems.

While the application of δ_e and δ_t modeling may be used to interpolate between sampling events during the growing season, the utility of the approach is severely limited during the dormant season, due to shifts in boundary layer conditions and evaporation dynamics at the soil surface. Additional experiments with the CGM using high temporal-resolution data collection systems similar to this study are warranted to reconcile discrepancies for CGM performance over changes in seasonality. While in this study, the assumption that T was negligible during the dormant season may be appropriate, this may not be the case in other systems.

6. Conclusions

This study demonstrated a novel methodology for partitioning ET using a combination of the high-resolution stable isotope, sap flux, and eddy covariance techniques over multiple seasons in a natural oak woodland site. To the best of our knowledge, this is the first report of utilizing a CRDS for ET partitioning at this timescale and under these conditions, and only the second report of direct partitioning comparisons between stable isotope derived f_T and f_T from sap flow and eddy covariance. Furthermore, we have developed predictive models for estimating the isotopic composition of the soil (δ_e) and xylem (δ_t) water (both components necessary for the stable isotope-based ET partitioning method) on days when field samples cannot be taken, by utilizing common environmental and micrometeorological data. With the high-frequency data made available through NEON, we found that δ^2H and $\delta^{18}O$ produced similar results, where, in the literature, δ^2H is typically only utilized. The utility of the stable isotope method for partitioning ET has been markedly improved with the with the integration of high-temporal resolution data collection systems, such as an EC tower coupled with a CRDS, along with field sampling of soil and twigs. However, further refinement of the methodology, particularly in natural systems, is still needed to reconcile potential biases inherent in this approach, and to test the utility of the proposed δ_e and δ_t models under wide-ranging conditions. We recommend that future investigations continue to refine the use of high frequency stable isotope data for novel application in partitioning ET, given the advantages of a single integrated whole-canopy estimation.

Supplementary Materials: The following are available online at http://www.mdpi.com/2073-4441/12/11/2967/s1, Figure S1: Distributed Plots Sapwood Area, Table S1: NEON Data Products Used in Study, Table S2: Sapwood Area to Basal Area, Table S3: Daily Keeling Plots.

Author Contributions: Conceptualization, G.W.M. and C.C.-N.; methodology, C.A., G.W.M., C.C.-N.; formal analysis, C.A.; resources, G.W.M., C.C.-N., C.A.; data curation, C.A.; writing—original draft preparation, C.A.; writing—review and editing, C.A., G.W.M., C.C.-N.; visualization, C.A.; supervision, G.W.M., C.C.-N.; project administration, C.A., G.W.M., C.C.-N., R.P.; funding acquisition, C.A., G.W.M., C.C.-N., R.P. All authors have read and agreed to the published version of the manuscript.

Funding: Graduate study was supported by the Texas A&M University College of Agriculture and Life Sciences Excellence Fellowship, as well as the Department of Ecosystem Science and Management George and Judy Dishman Endowed Graduate Fellowship. Funding for conferences, travel to field sites, and purchase of research materials were made possible in part by the Texas A&M University Office of Graduate and Professional Studies Research and Travel Award, the ESSM Graduate Student Travel Grant, and the ESSM Graduate Student Research Mini-Grant. The manuscript contents are solely the responsibility of the authors and do not necessarily represent the official views of the funding parties.

Acknowledgments: This research was conducted in partnership with the National Ecological Observatory Network (NEON) who allowed our team access to the research site and provided support services for this work. The authors would like to thank NEON for their provision of free, open access data to this project. Monika Kelley, Gary Henson, and Jarrett Jamison with Battelle Ecology provided technical support and maintenance help throughout the study. We would also like to thank Ajinkya Deshpande, Ashley Cross, Aaron Trimble, Miriam Catalan, and Manuel Flores for their assistance with sensor construction, site set-up, and field sampling.

Conflicts of Interest: The authors declare no conflict of interest.

References

1. Trenberth Kevin, E.; Smith, L.; Qian, T.; Dai, A.; Fasullo, J. Estimates of the Global Water Budget and Its Annual Cycle Using Observational and Model Data. *J. Hydrometeorol.* **2007**, *8*, 758. [CrossRef]
2. Wilcox, B.; Breshears, D.; Seyfried, M. Water Balance on Rangelands. *Encycl. Water Sci.* **2003**, 791–794. [CrossRef]

3.   Good, S.P.; Moore, G.W.; Miralles, D.G. A mesic maximum in biological water use demarcates biome sensitivity to aridity shifts. *Nat. Ecol. Evol.* **2017**, *1*, 1883–1888. [CrossRef]

4.   Helman, D.; Lensky, I.M.; Osem, Y.; Rohatyn, S.; Rotenberg, E.; Yakir, D. A biophysical approach using water deficit factor for daily estimations of evapotranspiration and CO_2 uptake in Mediterranean environments. *Biogeosciences* **2017**, *14*, 3909–3926. [CrossRef]

5.   Ford, C.R.; Hubbard, R.M.; Kloeppel, B.D.; Vose, J.M. A comparison of sap flux-based evapotranspiration estimates with catchment-scale water balance. *Agric. For. Meteorol.* **2007**, *145*, 176–185. [CrossRef]

6.   Shi, T.T.; Guan, D.X.; Wu, J.B.; Wang, A.Z.; Jin, C.J.; Han, S.J. Comparison of methods for estimating evapotranspiration rate of dry forest canopy: Eddy covariance, Bowen ratio energy balance, and Penman-Monteith equation. *J. Geophys. Res. Atmos.* **2008**, *113*, 15. [CrossRef]

7.   Tie, Q.; Hu, H.; Tian, F.; Holbrook, N.M. Comparing different methods for determining forest evapotranspiration and its components at multiple temporal scales. *Sci. Total Environ.* **2018**, *633*, 12–29. [CrossRef] [PubMed]

8.   Williams, D.G.; Cable, W.; Hultine, K.; Hoedjes, J.C.B.; Yepez, E.A.; Simonneaux, V.; Er-Raki, S.; Boulet, G.; de Bruin, H.A.R.; Chehbouni, A.; et al. Evapotranspiration components determined by stable isotope, sap flow and eddy covariance techniques. *Agric. For. Meteorol.* **2004**, *125*, 241–258. [CrossRef]

9.   Kool, D.; Agam, N.; Lazarovitch, N.; Heitman, J.L.; Sauer, T.J.; Ben-Gal, A. A review of approaches for evapotranspiration partitioning. *Agric. For. Meteorol.* **2014**, *184*, 56–70. [CrossRef]

10.   Aouade, G.; Ezzahar, J.; Amenzou, N.; Er-Raki, S.; Benkaddour, A.; Khabba, S.; Jarlan, L. Combining stable isotopes, Eddy Covariance system and meteorological measurements for partitioning evapotranspiration, of winter wheat, into soil evaporation and plant transpiration in a semi-arid region. *Agric. Water Manag.* **2016**, *177*, 181–192. [CrossRef]

11.   Xu, Z.; Yang, H.; Liu, F.; An, S.; Cui, J.; Wang, Z.; Liu, S. Partitioning evapotranspiration flux components in a subalpine shrubland based on stable isotopic measurements. *Bot. Stud.* **2008**, *49*, 351–361.

12.   Wilson, K.B.; Hanson, P.J.; Mulholland, P.J.; Baldocchi, D.D.; Wullschleger, S.D. A comparison of methods for determining forest evapotranspiration and its components: Sap-flow, soil water budget, eddy covariance and catchment water balance. *Agric. For. Meteorol.* **2001**, *106*, 153–168. [CrossRef]

13.   Moore, G.W.; Cleverly, J.R.; Owens, M.K. Nocturnal transpiration in riparian Tamarix thickets authenticated by sap flux, eddy covariance and leaf gas exchange measurements. *Tree Physiol.* **2008**, *28*, 521–528. [CrossRef]

14.   Granier, A. Evaluation of transpiration in a Douglas-fir stand by means of sap flow measurements. *Tree Physiol.* **1987**, *3*, 309–320. [CrossRef] [PubMed]

15.   Granier, A. Une nouvelle méthode pour la mesure du flux de sève brute dans le tronc des arbres. *Ann. Sci.* **1985**, *42*, 193–200. [CrossRef]

16.   Nier, A. The development of A high-resolution mass-spectrometer—A reminiscence. *J. Am. Soc. Mass Spectrom.* **1991**, *2*, 447–452. [CrossRef]

17.   Walker, C.D.; Brunel, J.P. Examining evapotranspiration in a semi-arid region using stable isotopes of hydrogen and oxygen. *J. Hydrol.* **1990**, *118*, 55–75. [CrossRef]

18.   Brunel, J.-P.; Walker, G.R.; Kennett-Smith, A.K. Field validation of isotopic procedures for determining sources of water used by plants in a semi-arid environment. *J. Hydrol.* **1995**, *167*, 351–368. [CrossRef]

19.   Brunel, J.P.; Walker, G.R.; Dighton, J.C.; Monteny, B. Use of stable isotopes of water to determine the origin of water used by the vegetation and to partition evapotranspiration. A case study from HAPEX-Sahel. *J. Hydrol.* **1997**, *188*, 466–481. [CrossRef]

20.   Munksgaard, N.C.; Wurster, C.M.; Bird, M.I. Continuous analysis of δ18O and δD values of water by diffusion sampling cavity ring-down spectrometry: A novel sampling device for unattended field monitoring of precipitation, ground and surface waters. *Rapid Commun. Mass Spectrom.* **2011**, *25*, 3706–3712. [CrossRef] [PubMed]

21.   Wei, Z.; Yoshimura, K.; Okazaki, A.; Kim, W.; Liu, Z.; Yokoi, M. Partitioning of evapotranspiration using high-frequency water vapor isotopic measurement over a rice paddy field. *Water Resour. Res.* **2015**, *51*, 3716–3729. [CrossRef]

22.   Soil Survey Staff, NRCS. *Web Soil Survey*; United States Department of Agriculture: Washington, DC, USA, 2019.

23.   NEON. *National Ecological Observatory Network*; Battelle: Boulder, CO, USA, 2020.

24. Michael, W.; Jan, H. Biohydrologic effects of eastern redcedar encroachment into grassland, Oklahoma, USA. *Biologia* **2013**, *68*, 1132–1135. [CrossRef]

25. Jones, H.G. *Plants and Microclimate: A Quantitative Approach to Environmental Plant Physiology*; Cambridge University Press: Cambridge, UK, 1992; p. 428.

26. Kljun, N.; Calanca, P.; Rotach, M.W.; Schmid, H.P. A simple two-dimensional parameterisation for Flux Footprint Prediction (FFP). *Geosci. Model Dev.* **2015**, *8*, 3695–3713. [CrossRef]

27. Venkatram, A. Estimating the Monin-Obukhov length in the stable boundary layer for dispersion calculations. *Bound. Layer Meteorol.* **1980**, *19*, 481–485. [CrossRef]

28. Zotarelli, L.; Dukes, M.D.; Romero, C.C.; Migliaccio, K.W.; Morgan, K.T. *Step by Step Calculation of the Penman-Monteith Evapotranspiration (FAO-56 Method)*; Institute of Food and Agricultural Sciences. University of Florida: Gainesville, FL, USA, 2018.

29. Moore, G.W.; Bond, B.J.; Jones, J.A.; Phillips, N.; Meinzer, F.C. Structural and compositional controls on transpiration in 40-and 450-year-old riparian forests in western Oregon, USA. *Tree Physiol.* **2004**, *24*, 481–491. [CrossRef] [PubMed]

30. Vertessy, R.A.; Watson, F.G.R.; O'Sullivan, S.K. Factors determining relations between stand age and catchment water balance in mountain ash forests. *For. Ecol. Manag.* **2001**, *143*, 13–26. [CrossRef]

31. McDowell, N.; Barnard, H.; Bond, B.J.; Hinckley, T.; Hubbard, R.M.; Ishii, H.; Köstner, B.; Magnani, F.; Marshall, J.D.; Meinzer, F.C.; et al. Relatsh between Tree Height Leaf Area: Sapwood Area Ratio. *Oecologia* **2002**, *132*, 12. [CrossRef]

32. Gebauer, T.; Horna, V.; Leuschner, C. Variability in radial sap flux density patterns and sapwood area among seven co-occurring temperate broad-leaved tree species. *Tree Physiol.* **2008**, *28*, 1821–1830. [CrossRef]

33. Clearwater, M.J.; Goldstein, G.; Holbrook, N.M.; Meinzer, F.C.; Andrade, J.L. Potential errors in measurement of nonuniform sap flow using heat dissipation probes. *Tree Physiol.* **1999**, *19*, 681–687. [CrossRef]

34. Mackay, D.S.; Ewers, B.E.; Loranty, M.M.; Kruger, E.L. On the representativeness of plot size and location for scaling transpiration from trees to a stand. *J. Geophys. Res. Biogeosci.* **2010**, *115*, 14. [CrossRef]

35. Xiao, W.; Wei, Z.; Wen, X. Evapotranspiration partitioning at the ecosystem scale using the stable isotope method—A review. *Agric. For. Meteorol.* **2018**, *263*, 346–361. [CrossRef]

36. Zhang, S.; Wen, X.; Wang, J.; Yu, G.; Sun, X. The use of stable isotopes to partition evapotranspiration fluxes into evaporation and transpiration. *Acta Ecol. Sin.* **2010**, *30*, 201–209. [CrossRef]

37. Barnes, C.J.; Allison, G.B. Tracing of water movement in the unsaturated zone using stable isotopes of hydrogen and oxygen. *J. Hydrol.* **1988**, *100*, 143–176. [CrossRef]

38. Sprenger, M.; Leistert, H.; Gimbel, K.; Weiler, M. Illuminating hydrological processes at the soil-vegetation-atmosphere interface with water stable isotopes. *Rev. Geophys.* **2016**, *54*, 674–704. [CrossRef]

39. Zimmermann, U.; Ehhalt, D.; Muennich, K.O. *Soil-Water Movement and Evapotranspiration: Changes in the Isotopic Composition of the Water*; International Atomic Energy Agency: Vienna, Austria, 1967.

40. Ehleringer, J.R.; Dawson, T.E. Water uptake by plants: Perspectives from stable isotope composition. *Plant Cell Environ.* **1992**, *15*, 1073–1082. [CrossRef]

41. Yepez, E.A.; Williams, D.G.; Scott, R.L.; Lin, G. Partitioning overstory and understory evapotranspiration in a semiarid savanna woodland from the isotopic composition of water vapor. *Agric. For. Meteorol.* **2003**, *119*, 53–68. [CrossRef]

42. Flanagan, L.B.; Comstock, J.P.; Ehleringer, J.R. Comparison of Modeled and Observed Environmental Influences on the Stable Oxygen and Hydrogen Isotope Composition of Leaf Water in *Phaseolus vulgaris* L. *Plant Physiol.* **1991**, *96*, 588. [CrossRef]

43. Keeling, C.D. The concentration and isotopic abundances of atmospheric carbon dioxide in rural areas. *Geochim. Cosmochim. Acta* **1958**, *13*, 322–334. [CrossRef]

44. Keeling, C.D. The concentration and isotopic abundances of carbon dioxide in rural and marine air. *Geochim. Cosmochim. Acta* **1961**, *24*, 277–298. [CrossRef]

45. Yakir, D.; Sternberg, L.d.S.L. The use of stable isotopes to study ecosystem gas exchange. *Oecologia* **2000**, *123*, 297–311. [CrossRef] [PubMed]

46. Craig, H.; Gordon, L.I. Deuterium and oxygen 18 variations in the ocean and the marine atmosphere. In Proceedings of the Conference on Stable Isotopes in Oceanographic Studies and Paleotemperatures, Spoleto, Italy, 26–30 July 1965; pp. 9–130.

47. Majoube, M. Fractionnement en oxygène 18 et en deutérium entre l'eau et sa vapeur. *J. De Chim. Phys.* **1971**, *68*, 1423–1436. [CrossRef]

48. Cappa, C.D.; Hendricks, M.B.; DePaolo, D.J.; Cohen, R.C. Isotopic fractionation of water during evaporation. *J. Geophys. Res. Atmos.* **2003**, *108*, 10. [CrossRef]

49. Merlivat, L. Molecular diffusivities of $H_2^{16}O$, $HD^{16}O$, and $H_2^{18}O$ in gases. *J. Chem. Phys.* **1978**, *69*, 2864–2871. [CrossRef]

50. Service, N.W. *Weather for Decatur*; National Oceanic and Atmospheric Administration: College Station, TX, USA, 2020.

51. Nehemy, M.F.; Benettin, P.; Asadollahi, M.; Pratt, D.; Rinaldo, A.; McDonnell, J.J. How plant water status drives tree source water partitioning. *Hydrol. Earth Syst. Sci. Discuss.* **2019**, *2019*, 1–26. [CrossRef]

52. Wilson, K.B.; Hanson, P.J.; Baldocchi, D.D. Factors controlling evaporation and energy partitioning beneath a deciduous forest over an annual cycle. *Agric. For. Meteorol.* **2000**, *102*, 83–103. [CrossRef]

Publisher's Note: MDPI stays neutral with regard to jurisdictional claims in published maps and institutional affiliations.

Article

Effects of Farming Activities on the Temporal and Spatial Changes of Hydrogen and Oxygen Isotopes Present in Groundwater in the Hani Rice Terraces, Southwest China

Chengjing Liu, Yuanmei Jiao *, Dongmei Zhao, Yinping Ding, Zhilin Liu and Qiue Xu

School of Tourism and Geographical Sciences, Yunnan Normal University, No. 768 Juxian Street, Chenggong District, Kunming 650500, China; 18487101130@163.com (C.L.); 18487365270@163.com (D.Z.); dyp520513@163.com (Y.D.); zhilin2015@foxmail.com (Z.L.); 15974668403@163.com (Q.X.)
* Correspondence: ymjiao@sina.com; Tel.: +86-1388-808-3272

Received: 16 December 2019; Accepted: 14 January 2020; Published: 17 January 2020

Abstract: Landform changes caused by human activities can directly affect the recharge of groundwater, and are reflected in the temporal and spatial changes in groundwater stable isotope composition. These changes are particularly evident in high-intensity farming areas. In this study, we tested and analyzed groundwater stable isotope samples at different elevations of rice terraces in a typical agricultural watershed of the Hani Terraces, a World Heritage Cultural Landscape in southwest China. Thus, we determined the characteristic variations and factors that influence the temporal and spatial effects on groundwater stable isotopes in the Hani Terraces, which are under the influence of high-intensity farming activities. The elevation gradients of δ^{18}O and δ^2H in groundwater are significantly increased due to farming activities. The values were 0.88‰ (100 m)$^{-1}$ and −4.5‰ (100 m)$^{-1}$, respectively, and they changed with time. The groundwater circulation cycle is approximately three months. We also used the special temporal and spatial variation characteristics of the groundwater isotopes as a way to evaluate the source and periodic changes of groundwater recharge. In addition, high-intensity rice farming activities, such as ploughing every year from October to January can increase the supply of terraced water to groundwater, thus ensuring the sustainability of rice cultivation in the terraces during the dry season. This demonstrates the role of human wisdom in the sustainable and benign transformation of surface cover and the regulation of groundwater circulation.

Keywords: Hani Rice Terraces; hydrogen and oxygen isotopes; groundwater recharge; elevation effect; deuterium excess

1. Introduction

Groundwater, as the world's largest freshwater resource, is critically important for irrigated agriculture, and hence for global food security [1]. The increasing use of groundwater resources for human consumption, agriculture, and industrial uses requires the protection of recharge areas from contamination or land use and land cover changes that may affect the quality and quantity of water in the aquifers [2–4]. Furthermore, climate change has the potential to exacerbate the problem in some regions. For example, it is predicted that anthropogenic climate change will affect rainfall amount and distribution in many parts of the world [5–8]. Under the influence of both climate change and human activities, depletion of water is widespread in large groundwater systems in both semi-arid and humid regions of the world [1]. Consequently, documenting the spatial patterns and determining the magnitude of groundwater recharge is important for understanding and managing groundwater systems using the appropriate tools [4].

One of the major changes that has occurred in earth and environmental sciences is the addition of rigorous analytical tools such as isotopes, to complement classical descriptive methods [9]. For example, the isotopic ratios of chemical elements such as oxygen and hydrogen of the water molecule ($^{18}O/^{16}O$ and $^{2}H/^{1}H$, respectively) are used as powerful tools to trace isotopes in hydrological processes [10–13]. Without any evaporation and exchange with dissolved gases or rocks, the stable isotopes (^{18}O and ^{2}H) can be considered as conservative, and thus reflect the mixture of the different recharge at the origin of the groundwater. Meanwhile, the abundance of ^{18}O and ^{2}H is driven by environmental conditions such as recharge altitude, recharge water source, and ambient temperature, which leads to a specific signature in the groundwater [14–16].

Since the 1970s, hydrogeologists have used stable water isotopes (^{2}H and ^{18}O) as tracers to explain the isotopic characteristics of groundwater across the seasons, and calculate groundwater recharge sources and the groundwater recharge altitude [17–19]. As research continued, groundwater recharge, the relationship between groundwater and surface water, the origin and evolution of groundwater, and groundwater quality gained increasing attention [20–23]. These studies have made a positive contribution towards understanding groundwater processes, protecting regional water resources, achieving regional water cycle stability, and maintaining the sustainability of social production and life.

Groundwater recharge in both moist and arid areas is susceptible to changes in land use/land cover, especially when this is due to human activities [24–26]. Previous studies have suggested that the global impact of land use and cover change (LUCC) on the hydrologic cycle may surpass that of recent climate change [27,28]. In Australia, removal of indigenous vegetation over the past 100 years has significantly increased groundwater recharge and caused a rise in the water-table [29]. Research on the Loess Plateau showed that groundwater recharge beneath natural sparse small-grass was 100 mm/year, but the conversion to winter wheat about 100 years ago reduced the groundwater recharge to 55 mm/year [26]. Therefore, the traditional recharge estimation method and isotope effect of groundwater has large errors in some areas.

This situation is particularly prominent in agricultural areas where humans have exerted large-scale changes on the terrestrial biosphere, primarily through agriculture. For example, Liu et al. (2016) studied the groundwater in the Malizhai River basin of the Hani Terraces and found that groundwater in the terrace area had different elevation gradients at different elevations [30]. Further analysis by Jiao et al. (2017) showed that the dominant landscape types at different elevations have a significant impact on the isotope composition of underground spring water [31]. This impact is mainly manifest in differences in the elevation gradients of stable isotopes present in spring water within the same landscape type and between different landscape types. All of the unique variations in the elevation effect on groundwater isotopes in this area point to the influence of changes in regional land use. At present, it is known that land use change can affect the infiltration of surface water. However, it is necessary to further analyze how land use change influences groundwater isotopes and their temporal and spatial distribution, especially the impact of the land use change caused by agricultural activities on regional groundwater circulation.

In this paper, we (1) analyzed the features of the elevation effect on the oxygen and hydrogen isotopic composition of groundwater across month and year; (2) analyzed and identified the influencing mechanisms of land use change caused by farming activities on the elevation gradient of groundwater isotopes from a spatial perspective; (3) estimated the recharge cycle and period of regional groundwater from a temporal perspective using the deuterium excess; and (4) deepened our understanding of how farming activities can improve groundwater recharge efficiency and maintain agricultural sustainability.

2. Study Area

The study area is located in the Bada area, the core area of the Hani Terraces, a World Heritage Cultural Landscape in Yuanyang County, Yunnan Province. It belongs to the Quanfuzhuang catchment of the upstream tributary of the Malizhai River basin, located on the northern slope of the Southern Ailao Mountains, with the longitude ranging from 102°43′16″ to 102°50′39″ E, and the latitude ranging from 23°5′20″ to 23°13′18″ N (Figure 1). The terrain in the whole catchment is high in the north and low in the south. The catchment is fan-shaped from north to south, covering an area of 13.92 km^2. It accounts for 16.73% of the total area of the Malizhai River basin (83.2 km^2). The elevation of the catchment ranges from 1475 m to 2261 m. The elevation of the terrace area within the catchment ranges from 1475 m to 1737 m. The terrace area is located in the middle and lower reaches of the catchment and accounts for approximately 40% of the total area of the catchment. The terrain is characterized by a change from gentle slopes to steep slopes, with steepness mainly >25°. In terms of the climate, the annual precipitation is 1532.19 mm, and the annual evaporation is 1500.60 mm. The distribution of precipitation in the dry and rainy seasons is uneven. The rainy season is generally considered to be from May to October, and the dry season is from November to April. The precipitation in the rainy season accounts for 72% and the dry season precipitation accounts for 28%. The entire catchment belongs to the headwater area of the Malizhai River, a first-order tributary on the right bank of the Yuanjiang River in the upper reaches of the Honghe River. Abundant rainfall in the catchment provides sufficient water for the river. The drainage system of the catchment develops in the form of tree branches. There are two primary tributaries in the east and the west, which merge into the Malizhai River at the lowest part of the catchment. In addition, due to the impact of the fault structures and deep cutting terrain, the bedrock fissure water in the area mostly migrates from high to low water levels. It is discharged in the form of springs or diffused flow in gullies and foothills, according to the characteristics of recharge and discharge in the vicinity. In the study area, a large amount of groundwater is discharged into springs.

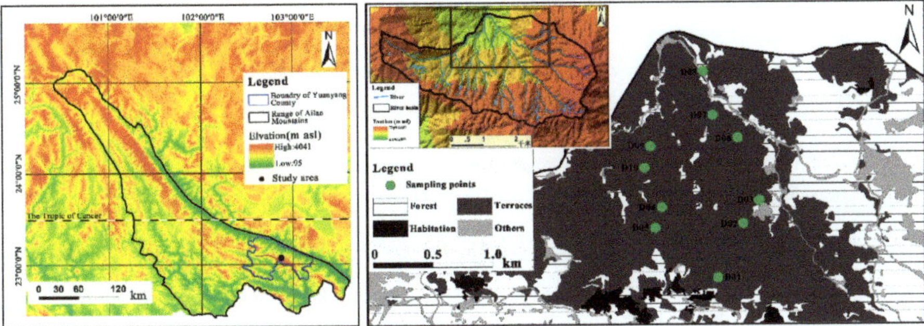

Figure 1. Study area and distribution of sampling points.

According to field investigation and the 1:200,000 regional geological report of the Yuanyang area [32], there are two types of aquifer: pore aquifer and fissured aquifer, in which the pore aquifer includes Qualified Design Listing (Qdl) and Qualified Eluvium Listing (Qel). The thickness of two strata with weak water abundance is between 0.1 m and 10 m. The fissured aquifer is gneiss and granulite in the lower part of the formation, with developed cracks and joint. There is a network of cracks in the weathering zone of these rocks. The difference in water abundance is controlled by the degree of development of cracks and joints. As a whole, the water abundance is better. Moreover, the water permeability of the rock with undeveloped joints and cracks is poor, forming an aquiclude. Precipitation is the most important source of groundwater recharge. The runoff discharge and buried depth of groundwater are obviously controlled by the terrain. Generally, groundwater has a short recharge path, with on-site recharge and on-site discharge. However, the groundwater is also recharged with the surface water in the terrace area and the lateral infiltration water of the upper terrace and the lower terrace, but the overall recharge is weak. In terms of the hydrochemical characteristics of the water in the River basin, the groundwater in the rainy season is classified as HCO$_3$-Ca with a pH value of 7.76 and total hardness of 81.54. The groundwater in the dry season is HCO$_3$-Ca·Na with a pH value of 6.89 and total hardness of 7.46. The ion content of the groundwater varies greatly in the dry and rainy seasons, which indicates that the recharge sources of groundwater are different in different seasons.

3. Data Sources and Research Methods

3.1. Sample Collection and Testing

Remote sensing image interpretation and digital elevation model analysis were conducted in the study area. Based on the results of our field investigation and taking into account water isotope sampling standards and precautions, sampling points were arranged along different elevation gradients in the Quanfuzhuang catchment of the Hani Terraces. The elevation of the sampling points ranged from 1500 m to 2260 m, and the difference in elevation was 760 m (Figure 1). Samples were collected for one year from May 2015 to April 2016, and the time of alternation between dry and rainy seasons was kept consistent in order to comprehensively observe the change in isotope composition of samples in a dry-rainy season cycle. Ten groundwater sampling points were established and 120 groundwater samples were obtained. All the groundwater samples are spring water with a slow flow rate (0~0.2 m/s).

The hydrogen and oxygen isotope tests were conducted in the Key Laboratory of Plateau Lake Ecology and Global Change, Yunnan Normal University. Using a Picarro L2130-i ultra-high precision liquid water and moisture isotope analyzer, the measurement accuracy of $\delta^{18}O$ and $\delta^{2}H$ was ±0.1‰ and ±0.5‰, respectively. The final analysis results were expressed by the thousandth difference relative to the Vienna standard mean ocean water (V-SMOW) as follows:

$$\delta^{2}H = \left(\frac{R_{D-sample}}{R_{V-SMOW}} - 1\right) \times 1000‰ \tag{1}$$

$$\delta^{18}O = \left(\frac{R_{O-sample}}{R_{V-SMOW}} - 1\right) \times 1000‰ \tag{2}$$

where R_{sample} and R_{V-SMOW} represent the oxygen and hydrogen stable isotope ratios of R ($^{18}O/^{16}O$) and R (D/H) in the water samples and *V-SMOW*, respectively.

3.2. Analysis Methods and Data Sources

The analysis methods included SPSS statistical analysis, correlation testing, ArcGIS geostatistical analysis, and cartographic generalization. The meteorological data was obtained from the small meteorological observation station located in Quanfuzhuang in the middle of the basin, which automatically recorded the hourly data for temperature, humidity, precipitation, and other relevant meteorological elements. The remote sensing images of the study area, showing the terrain and location information of sampling points were downloaded from the Local Space Viewer software at a resolution of 0.6 m. The digital elevation model was provided by the National Geospatial Data Cloud. The hydrogeological data of the basin was derived from the field investigation of the study area and the 1:200,000 regional geological report of the Yuanyang area.

4. Results and Analysis

4.1. Compositional Characteristics of Stable Isotopes Present in Groundwater

According to the analysis of the test results for stable isotopes present in 120 groundwater samples during the one-year study period, in the middle-altitude terraces of the Quanfuzhuang catchment of the Hani Terraces, the $\delta^{18}O$ values of groundwater ranged from −9.65‰ to −3.01‰, with an average value of −7.29‰, and the $\delta^{2}H$ values ranged from −65.86‰ to −37.60‰, with an average value of −51.32‰. The $\delta^{18}O$ and $\delta^{2}H$ values of groundwater isotopes during the rainy season ranged from −9.65‰ to −3.01‰ and from −65.86‰ to −37.60‰, with average values of −7.10‰ and −50.52‰, respectively. During the dry season, the value ranges were from −9.55‰ to −6.01‰ and from −62.38‰

to −43.37‰, with average values of −7.47‰ and 52.11‰, respectively. Overall, there were minor differences between the dry and rainy season. The average $\delta^{18}O$ and δ^2H values during the rainy season were slightly higher than those during the dry season by 0.37‰ and 1.59‰, respectively. Both the maximum and minimum values of $\delta^{18}O$ and δ^2H occurred during the rainy season. Spatially, the groundwater stable isotope values ($\delta^{18}O$ and δ^2H) along the elevation gradient of 300 m in the terrace area gradually decreased with increasing elevation. The differences between the stable isotope values at the highest and the lowest sampling points were −3.04‰ and −15.28‰, respectively, and the stable isotope values of spring water changed significantly with elevation.

Based on the measured values of hydrogen and oxygen isotopes present in groundwater from May 2015 to April 2016, the simple linear regression analysis of δ^2H and $\delta^{18}O$ in groundwater showed that annually, the relationship between δ^2H and $\delta^{18}O$ in the groundwater in the study area is $\delta^2H = 4.98\delta^{18}O - 15.01$ ($R^2 = 0.92$, n = 144). In the rainy season, $\delta^2H = 4.89\delta^{18}O - 15.79$ ($R^2 = 0.90$, n = 72); in the dry season, $\delta^2H = 5.19\delta^{18}O - 13.31$ ($R^2 = 0.95$, n = 72) (Figure 2). The slopes of the δ^2H–$\delta^{18}O$ correlation lines of groundwater (LGWL) in the study area were similar for the whole year, the rainy season, and the dry season. The slope of the rainy season was equal to that of the whole year, while the slope of the dry season was slightly larger than that of the rainy season. In addition, the slopes for the whole year, the rainy season, and the dry season in the study area were less than the measured slope (8.31) of the annual local meteoric water line (LMWL) [33] and the slope (8.14 ± 0.02) of the global meteoric water line (GMWL) [34], which indicates that the main groundwater recharge source is terraced water with strong evaporation.

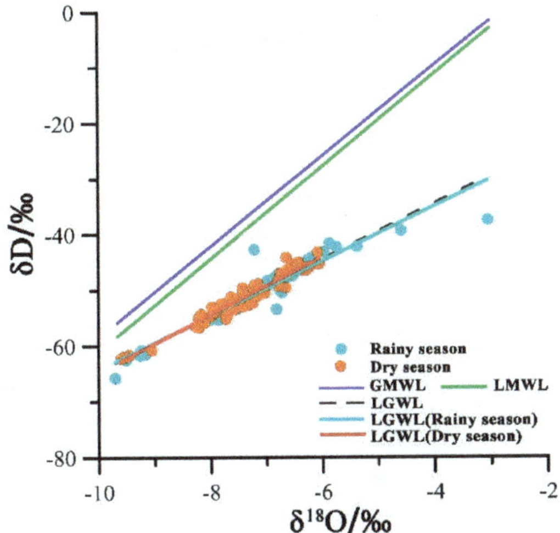

Figure 2. Local groundwater line (LGWL) for the study area across season, based on the data of individual sampling from May 2015 to April 2016.

4.2. Variation in the Elevation Effect on the Stable Isotope Values of Groundwater

The groundwater isotope values ($\delta^{18}O$ and δ^2H) of 10 sampling points (D_{01}–D_{10}) in the study area were averaged for each month in one year. It was found that the spatial distribution of the annual spring water isotope values demonstrated an overall decrease with increasing elevation (Figure 3). Through the correlation analysis of the elevation effect on the annual average stable isotope values ($\delta^{18}O$ and δ^2H) of groundwater samples, it was found that the correlation coefficients (r_α) for the annual groundwater isotope values ($\delta^{18}O$ and δ^2H) and elevation were −0.80 and −0.78, respectively, at a significance level of 0.05. The annual stable isotope values ($\delta^{18}O$ and δ^2H) of groundwater have significant correlations with the elevation (n = 10, r_a = 0.63). In other words, there are significant elevation effects. The regression equation can be obtained by regression analysis of the annual average values of $\delta^{18}O$ and δ^2H in groundwater and the elevations:

$$\delta^{18}O = -0.0088\,h + 6.91 \ (R^2 = 0.64, n = 10) \tag{3}$$

$$\delta^2H = -0.0450\,h + 21.27 \ (R^2 = 0.60, n = 10) \tag{4}$$

where h is the elevation, and the unit is m. It was found that the elevation gradients of the annual stable isotope values ($\delta^{18}O$ and δ^2H) of groundwater in the study area were $-0.88‰$ $(100\ m)^{-1}$ and $-4.5‰$ $(100\ m)^{-1}$, respectively.

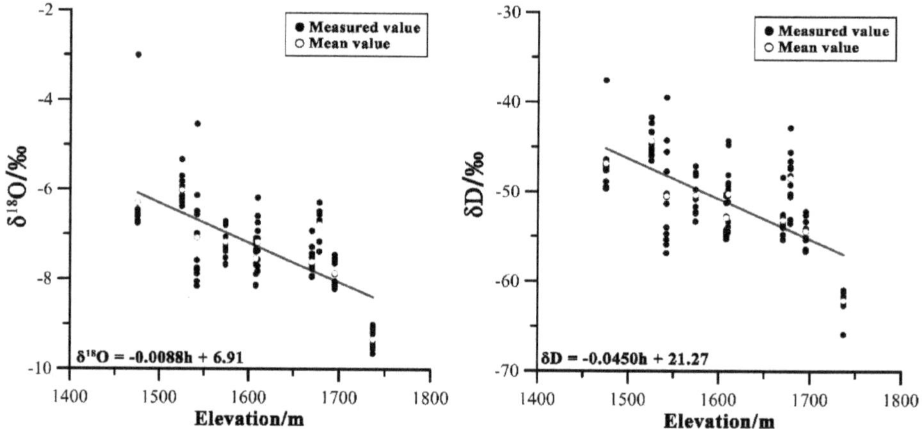

Figure 3. Variation in groundwater isotope values ($\delta^{18}O$ and δ^2H) with elevation.

In order to further explore the temporal change in elevation effect in the study area, we conducted correlation analysis and regression analysis of the groundwater stable isotope values ($\delta^{18}O$ and δ^2H) of 10 sampling points (D_{01}–D_{10}) with elevation. It was found that the groundwater stable isotope values in both the dry and rainy season in the study area had significant correlation with elevation (Table 1). Therefore, the effects of isotopic elevation on the groundwater in the study area were clearly demonstrated. The elevation gradients of the groundwater stable isotope values ($\delta^{18}O$ and δ^2H) in the dry season were $-0.74‰$ $(100\ m)^{-1}$ and $-3.76‰$ $(100\ m)^{-1}$, respectively, which were lower than those of the spring water stable isotope values ($-1.02‰$ $(100\ m)^{-1}$ and $-5.25‰$ $(100\ m)^{-1}$, respectively) in the rainy season, indicating that there was a higher isotopic elevation gradient in the rainy season.

Table 1. The relationship between the stable isotope value of groundwater and the elevation in different months in the study area.

Month	$\delta^{18}O$		$\delta^{2}H$	
	Correlation Coefficient(r_α)	Regression Equation	Correlation Coefficient(r_α)	Regression Equation
May	−0.70 *	$\delta^{18}O = -0.0073\,h + 4.64$	−0.76 *	$\delta^{2}H = -0.0452\,h + 21.73$
Jun	−0.77 *	$\delta^{18}O = -0.0084\,h + 6.59$	−0.80 *	$\delta^{2}H = -0.0491\,h + 29.92$
Jul	−0.75 *	$\delta^{18}O = -0.0109\,h + 10.60$	−0.81 *	$\delta^{2}H = -0.0583\,h + 44.24$
Aug	−0.85 *	$\delta^{18}O = -0.0170\,h + 20.46$	−0.89 *	$\delta^{2}H = -0.0741\,h + 69.91$
Sep	−0.67 *	$\delta^{18}O = -0.0077\,h + 5.06$	−0.64 *	$\delta^{2}H = -0.0461\,h + 22.29$
Oct	−0.76 *	$\delta^{18}O = -0.0099\,h + 8.51$	−0.55	$\delta^{2}H = -0.0420\,h + 16.05$
Nov	−0.56	$\delta^{18}O = -0.0066\,h + 3.14$	−0.56	$\delta^{2}H = -0.0336\,h + 1.48$
Dec	−0.66 *	$\delta^{18}O = -0.0074\,h + 4.36$	−0.63 *	$\delta^{2}H = -0.0379\,h + 8.39$
Jan	−0.58	$\delta^{18}O = -0.0069\,h + 3.64$	−0.53	$\delta^{2}H = -0.0360\,h + 6.66$
Fed	−0.70 *	$\delta^{18}O = -0.0081\,h + 5.56$	−0.71 *	$\delta^{2}H = -0.0448\,h + 20.97$
Mar	−0.76 *	$\delta^{18}O = -0.0083\,h + 6.10$	−0.66 *	$\delta^{2}H = -0.0355\,h + 5.02$
Apr	−0.83 *	$\delta^{18}O = -0.0072\,h + 4.27$	−0.83 *	$\delta^{2}H = -0.0380\,h + 8.60$
Rainy season	−0.87 *	$\delta^{18}O = -0.0102\,h + 9.31$	−0.84 *	$\delta^{2}H = -0.0525\,h + 34.02$
Dry season	−0.70 *	$\delta^{18}O = -0.0074\,h + 4.51$	−0.67 *	$\delta^{2}H = -0.0376\,h + 8.52$
Annual	−0.80 *	$\delta^{18}O = -0.0088\,h + 6.91$	−0.78 *	$\delta^{2}H = -0.0450\,h + 21.27$

Note: the sample number is 10. When $|r_\alpha| \geq 0.63$, the correlation coefficient passes the correlation test, and this is denoted as "*".

During the dry-rainy season cycle (12 months) in the study area, most months demonstrated significant elevation effects, except for November and January which showed no elevation effect was shown (Table 2). The months with elevation effects accounted for 83.3% of the dry-rainy season cycle (12 months). Further analysis showed that November and January, when there was no elevation effect, and December, when the elevation effect was not significant, were in the transition period from the rainy season (May to October) to the dry season (November to April of the following year). The change in precipitation during this period may be one of the factors that affects groundwater elevation effects in the study area. The impact of farming activities in this area is also an important factor. Therefore, the change in elevation effect on groundwater during the transition period between rainy and dry seasons is discussed further.

Table 2. $\delta^{18}O$ and δ^2H of groundwater at sampling points along the elevation transect.

Month	#	1	2	3	4	5	6	7	8	9	10
	Elevation/m	1737	1678	1610	1670	1695	1542	1525	1475	1574	1608
	Longitude/E	102°45.534′	102°45.627′	102°45.685′	102°45.315′	102°45.291′	102°45.602′	102°45.498′	102°45.469′	102°45.269′	102°45.759′
	Latitude/N	23°06.367′	23°06.563′	23°06.647′	23°06.618′	23°06.543′	23°06.873′	23°06.944′	23°07.110′	23°06.837′	23°06.759′
May	$\delta^{18}O/‰$	−9.12	−6.77	−6.19	−7.43	−7.58	−6.57	−6.14	−6.76	−7.06	−7.39
	$\delta^2H/‰$	−61.58	−53.49	−44.73	−52.79	−53.28	−47.70	−45.07	−48.86	−50.55	−52.71
June	$\delta^{18}O/‰$	−9.07	−6.68	−6.74	−6.92	−7.45	−6.14	−5.70	−6.45	−6.79	−7.14
	$\delta^2H/‰$	−61.14	−50.47	−48.08	−48.33	−52.21	−44.26	−42.35	−46.85	−47.83	−50.43
July	$\delta^{18}O/‰$	−9.21	−6.71	−6.92	−7.65	−7.52	−4.54	−5.92	−6.48	−6.79	−7.09
	$\delta^2H/‰$	−61.80	−50.20	−49.24	−53.92	−52.58	−39.50	−43.28	−47.44	−48.19	−50.46
August	$\delta^{18}O/‰$	−9.19	−6.55	−7.82	−7.77	−8.04	−6.49	−5.82	−3.01	−6.72	−7.33
	$\delta^2H/‰$	−60.94	−47.34	−54.48	−53.43	−55.35	−45.52	−41.69	−37.60	−47.12	−51.23
September	$\delta^{18}O/‰$	−9.65	−6.74	−7.70	−7.49	−7.89	−7.75	−6.20	−6.70	−6.71	−7.13
	$\delta^2H/‰$	−65.86	−47.23	−54.42	−53.55	−54.33	−54.70	−44.64	−47.64	−48.01	−50.21
October	$\delta^{18}O/‰$	−9.46	−7.16	−7.56	−7.94	−8.17	−7.80	−5.33	−6.46	−7.30	−7.22
	$\delta^2H/‰$	−62.63	−42.83	−53.31	−55.29	−56.36	−55.35	−42.34	−46.76	−50.81	−51.19
November	$\delta^{18}O/‰$	−9.40	−6.29	−7.83	−7.72	−8.06	−8.16	−6.17	−6.54	−7.52	−7.66
	$\delta^2H‰$	−61.80	−46.53	−53.91	−53.95	−55.03	−56.85	−45.85	−46.96	−51.67	−54.51
December	$\delta^{18}O/‰$	−9.47	−6.74	−7.10	−7.94	−8.21	−8.05	−6.29	−6.58	−7.68	−7.89
	$\delta^2H/‰$	−62.06	−48.15	−48.98	−54.77	−56.59	−55.89	−45.98	−47.34	−53.31	−54.23
January	$\delta^{18}O/‰$	−9.50	−6.49	−6.60	−7.93	−8.17	−7.88	−6.38	−6.70	−7.66	−8.15
	$\delta^2H/‰$	−62.04	−45.55	−44.30	−53.08	−55.10	−53.97	−45.14	−47.10	−52.35	−55.05
February	$\delta^{18}O/‰$	−9.46	−6.69	−7.18	−7.76	−8.12	−7.58	−6.03	−6.57	−7.37	−8.13
	$\delta^2H/‰$	−62.11	−47.16	−49.72	−52.56	−54.41	−51.34	−43.37	−46.39	−49.71	−54.86
March	$\delta^{18}O/‰$	−9.55	−7.15	−7.37	−7.27	−7.57	−6.98	−6.01	−6.61	−7.19	−7.68
	$\delta^2H/‰$	−62.38	−49.19	−50.35	−52.79	−53.31	−50.59	−45.45	−49.66	−52.42	−55.25
April	$\delta^{18}O/‰$	−9.01	−7.38	−7.10	−7.52	−7.64	−6.99	−6.24	−6.73	−7.23	−7.65
	$\delta^2H/‰$	−60.92	−53.06	−51.07	−53.53	−54.05	−50.17	−46.50	−49.41	−52.27	−54.62

5. Discussion

5.1. Spatial Effect: The Influence of Terraced Farming on the Elevation Effect on Hydrogen and Oxygen Isotopes Present in Groundwater

It is generally believed that the aquifer in the study area was dug through due to large-scale terrace excavation in the area. Consequently, a large amount of groundwater was discharged to the surface in the terrace area, resulting in the mixing of groundwater and surface water and reducing the elevation effect on groundwater in the study area. However, the current research indicates that groundwater in the terrace area still has a significant elevation effect, which is clearly related to the fact that the terraces in the study area are rice terraces.

Firstly, groundwater in the terrace area will be recharged by the terraced water and lateral infiltration caused by the height differences between the terraces. However, the overall influence is weak. This is the reason why the overall elevation effect on groundwater in the study area was relatively significant (Figure 4). Research on the Hani Terraces by Jiao (2009) also showed that the Hani Terraces are irrigated throughout the year with the exception of the harvest season (around October). The water depth is generally between 20 cm and 25 cm and the terrace water storage is 0.25 m^3/m^2 [35]. The bottom mud of the terraces is generally clay with low permeability, and farmers conduct real-time maintenance. Man-made rice terraces have better water-holding capacity, therefore, the loss of field water caused by leakage in terraces is almost negligible [35]. Rice cultivation over a long time and regular repair to prevent terrace leakage make surface water (field water) infiltration in the terraces quite difficult; this is the key reason that terrace groundwater maintains a better elevation gradient effect.

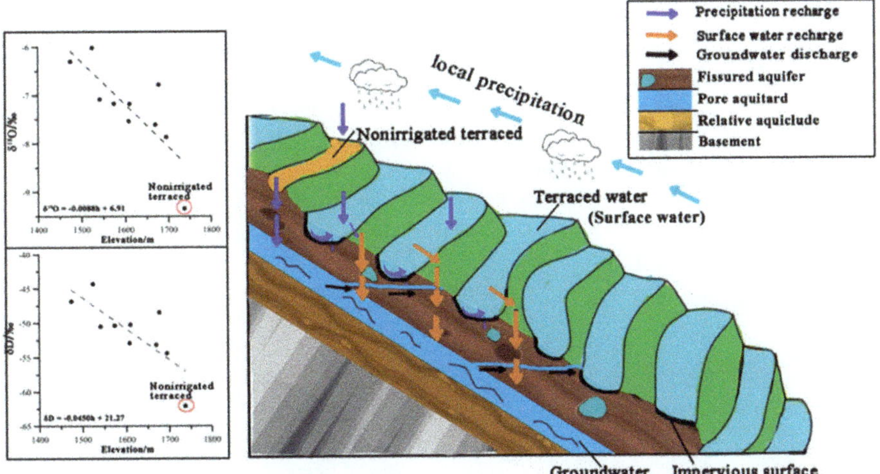

Figure 4. Spatial effect on groundwater under the influence of farming activities (schematic diagram).

However, the elevation gradients of the annual stable isotope values ($\delta^{18}O$ and δ^2H) of groundwater in the terrace area are −0.88‰ (100 m)$^{-1}$ and −4.5‰ (100 m)$^{-1}$, respectively. These values are more than twice the elevation effect gradients of the stable isotope values ($\delta^{18}O$ and δ^2H) of groundwater in Slovenia: −0.33‰ (100 m)$^{-1}$ (r = −0.89, $p < 0.01$) and −2.40‰ (100 m)$^{-1}$ (r = −0.89, $p < 0.01$), respectively [36]. Compared with the elevation gradient of $\delta^{18}O$ of the global precipitation, which is between −0.5‰ (100 m)$^{-1}$ and −0.15‰ (100 m)$^{-1}$ with an average value of −0.28‰ (100 m)$^{-1}$ [37], the elevation gradients of groundwater isotopes in the study area completely exceed this range. However, the elevation gradient of $\delta^{18}O$ of groundwater in the mountainous Hasi area in northwest China is −0.118‰ (100 m)$^{-1}$, which is less than this range [38]. This is due to the influence of farming activities on the surface water infiltration rate. As shown in Figure 4, the infiltration rate of rice terraces is different to that of dry fields. The surface water in rice terraces is controlled by the impermeable layer at the bottom of the terraces, which means water cannot infiltrate rapidly into the soil. Due to evaporation on the surface over a long period of time, the water isotope value tends to be more positive. In dry fields, precipitation and surface water can infiltrate rapidly into soil, so the isotope value tends to be more negative. When water in two different regions is recharged to the groundwater, if the elevation of the dry field is higher, the spatial difference in isotope values will be intensified and the elevation gradient of the groundwater stable isotope values will be larger. If the elevation of the dry field is lower, the spatial difference will be weakened, and the elevation gradient will be smaller or even tend to be positive. Therefore, the groundwater elevation effect gradient can be used as an indicator to determine the source of groundwater recharge, which is closely related to the landform change caused by local farming activities, and there are large differences between different regions.

5.2. Temporal Effect: The Relationship between the Terrace Farming Cycle and the Groundwater Recharge Cycle

The deuterium excess (also called d-excess, or *d* value) was first proposed and defined as $d = \delta^2H - 8\delta^{18}O$ by Dansgaared [39]. It reflects the degree of deviation of the composition of the hydrogen and oxygen isotope from the global meteoric water line, and to a certain extent, the differences between different water vapor sources. In detailed studies of the *d* value, many researchers have also applied the *d* value to other water bodies for analysis. In order to clarify the temporal variations in the elevation effect on groundwater stable isotopes in the study area, the *d* value of the groundwater was analyzed. As shown in Figure 5, the variation in the deuterium excess value of groundwater in the study area in each month was between 5.72‰ and 8.99‰, with an average value of 6.98‰, and the *d* value had significantly different characteristics in terms of time. In particular, the *d* value from May to August was relatively low, ranging from 5.4‰ to 6‰. From September to February, the *d* value increased continuously and was between 7‰ and 9‰; and from March to April it dropped below 7‰. From September to February in the study area, the change in *d* value of spring water was relatively stable, and this coincided with a period of precipitation during the rainy season, which to some extent indicates that the groundwater in this period was likely recharged by the infiltration of water bodies by precipitation and surface water. This was three months after the rainy season in the study area. Therefore, it takes approximately three months for precipitation to infiltrate into the groundwater layer in the study area.

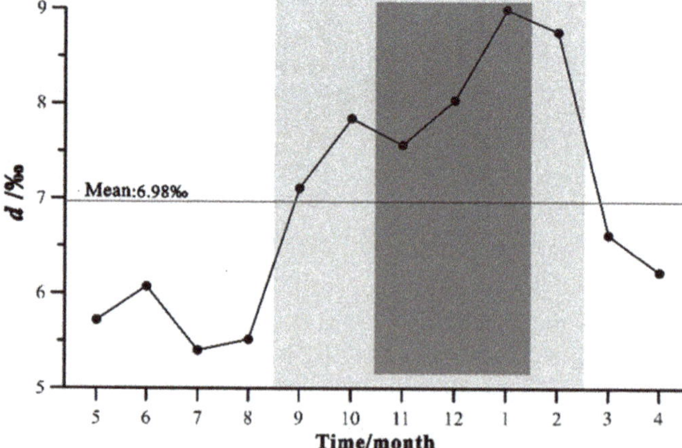

Figure 5. Variation in the deuterium excess of groundwater with time in the study area.

In the winter after rice harvesting, the Hani people need to plough the terraces to repair them. This is the reason why the elevation effect on groundwater in the study area was not significant in some months. The results of field investigation and research by local cultural researchers shows that local residents in the study area are engaged in large-scale agricultural activities related to terrace farming between November and January. This mainly involves the deep ploughing of terraces and re-irrigation (Figure 6). During this period, the impervious layer originally formed in the terraces is destroyed by human activity. Consequently, a large amount of field water or other surface water, for example, ditch water is recharged to the groundwater layer, which strengthens the supply of surface water to groundwater and thus weakens the elevation effect on groundwater in this period. The rapid increase in the *d* value of groundwater in the study area from December to January also indicates that there is a significant change in the supply water source during this period.

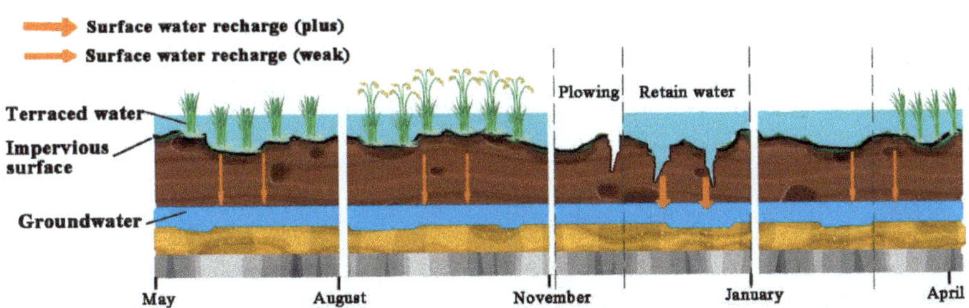

Figure 6. Spatial effect on groundwater under the influence of farming activities.

The ploughing activities are necessary for the maintenance of the terraces. Such activities also accelerate the infiltration of terraced water, thus increasing the supply of terraced water to groundwater. This provides sufficient water for groundwater recharge in the dry season and lays a solid foundation for farming in the early dry season of the following year. This is one of the key factors for sustainable cultivation in the Hani Terraces.

6. Conclusions

Through the testing and analysis of groundwater stable isotope samples at different elevations, this study has revealed the characteristic variations and influencing factors of temporal and spatial effects on the stable isotopes of groundwater in the Hani Terraces, which are under the influence of high-intensity farming activities. It was found that moderate farming activities can increase infiltration into groundwater. We also used the special temporal and spatial variation characteristics of groundwater isotopes as a means to predict the source and periodic changes in groundwater recharge.

In terms of spatial effects, the degree of mixing between groundwater and surface water in rice terraces maintained by farming activities is small, unlike dryland terraces. Stable isotopes of groundwater in the terraces have significant elevation effects. The elevation gradients of $\delta^{18}O$ and δ^2H were $-0.88‰$ $(100\ m)^{-1}$ and $-4.5‰$ $(100\ m)^{-1}$, respectively. The values in the terrace area exceeded the elevation gradients of atmospheric precipitation. This indicates that the multi-source groundwater recharge and farming activities in the study area create uncertainty about the location of surface water infiltration, which increases or decreases the elevation gradient of regional groundwater. The elevation gradient of groundwater can be used as a key index for assessing groundwater recharge.

In terms of temporal effects, rice cultivation is the key factor influencing the change in elevation effect on groundwater in the terrace area. The high-intensity rice cultivation that takes place from October to January accelerates the infiltration of surface water and strengthens the interaction between groundwater and surface water, resulting in no obvious elevation effect in this period. According to the temporal variation in the deuterium excess value of groundwater in the study area, it was inferred that the groundwater circulation time in the terrace area is approximately three months, and the sudden increase in the *d* value from October to January is also related to high-intensity rice cultivation.

In addition, high-intensity rice cultivation during the period from October to January increases the supply of terraced water to groundwater. This provides sufficient water for groundwater recharge before the dry season and thus ensures the sustainability of rice cultivation activities in the terraces during the dry season. This is one of the key reasons why the Hani Terraces have lasted for more than 1300 years, and also reflects how human wisdom can play a role in the sustainable and benign transformation of surface cover and the regulation of groundwater circulation.

Author Contributions: Conceptualization, Y.J.; Funding acquisition, Y.J.; Writing—original draft, C.L.; —review & editing, Y.J.; Formal analysis, D.Z. and Q.X.; Software, Y.D.; Investigation, Z.L.; Resources, Y.J. and Q.X. All authors have read and agreed to the published version of the manuscript.

Funding: This research was funded by the National Natural Science Foundation of China grant number 41271203 and 41761115. The APC was funded by School of Tourism and Geographical Sciences, Yunnan Normal University.

Acknowledgments: The authors thank Hucai Zhang of Yunnan University for the lab experiment with the water samples, Mingjun Zhang and Master's student Xiuxiu Yu of Northwest Normal University, and two anonymous reviewers and the editorial staff for constructive feedback that resulted in an improved manuscript.

Conflicts of Interest: The authors declare no conflicts of interest. The funders had no role in the design of the study; in the collection, analyses, or interpretation of data; in the writing of the manuscript; or in the decision to publish the results.

References

1. Aeschbachhertig, W.; Gleeson, T. Regional strategies for the accelerating global problem of groundwater depletion. *Nat. Geosci.* **2012**, *5*, 853–861. [CrossRef]

2. Michael, I.; Rizzo, L.; McArdell, C.S.; Manaia, C.M.; Merlin, C.; Schwartz, T.; Dagot, C.; Fatta-Kassinos, D. Urban wastewater treatment plants as hotspots for the release of antibiotics in the environment: A review. *Water Res.* **2013**, *47*, 957–995. [CrossRef] [PubMed]

3. Gleeson, T.; Befus, K.M.; Jasechko, S.; Luijendijk, E.; Cardenas, M.B. The global volume and distribution of modern groundwater. *Nat. Geosci.* **2016**, *9*, 161–164. [CrossRef]

4. Cartwright, I.; Cendon, D.I.; Currell, M.; Meredith, K.T. A review of radioactive isotopes and other residence time tracers in understanding groundwater recharge: Possibilities, challenges, and limitations. *J. Hydrol.* **2017**, *555*, 797–811. [CrossRef]

5. Lei, Y.; Hoskins, B.J.; Slingo, J.M. Exploring the interplay between natural decadal variability and anthropogenic climate change in summer rainfall over China. Part I: Observational evidence. *J. Clim.* **2011**, *24*, 4584–4599. [CrossRef]

6. Meixner, T.; Manning, A.H.; Stonestrom, D.A.; Allen, D.M.; Ajami, H.; Blasch, K.W.; Brookfield, A.E.; Castro, C.L.; Clark, J.F.; Gochis, D.J.; et al. Implications of projected climate change for groundwater recharge in the western United States. *J. Hydrol.* **2016**, *534*, 124–138. [CrossRef]

7. Rimi, R.H.; Haustein, K.; Allen, M.R.; Barbour, E. Risks of pre-monsoon extreme rainfall events of Bangladesh: Is anthropogenic climate change playing a role? *Bull. Am. Meteorol. Soc.* **2019**, *100*, S61–S65. [CrossRef]

8. Roderick, T.P.; Wasko, C.; Sharma, A. Atmospheric moisture measurements explain increases in tropical rainfall extremes. *Geophys. Res. Lett.* **2019**, *46*, 1375–1382. [CrossRef]

9. Negrel, P.; Ollivier, P.; Flehoc, C.; Hube, D. An innovative application of stable isotopes (δ^2H and δ^{18}O) for tracing pollutant plumes in groundwater. *Sci. Total Environ.* **2017**, *578*, 495–501. [CrossRef]

10. Dansgaared, W. The abundance of O^{18} in atmospheric water and water vapor. *Tellus* **1953**, *5*, 461–469.

11. Gat, J.R. Oxygen and hydrogen isotopes in the hydrologic cycle. *Annu. Rev. Earth Planet. Sci.* **1996**, *24*, 225–262. [CrossRef]

12. Juras, R.; Pavlasek, J.; Vitvar, T.; Sanda, M.; Holub, J.; Jankovec, J.; Linda, M. Isotopic tracing of the outflow during artificial rain-on-snow event. *J. Hydrol.* **2016**, *541*, 1145–1154. [CrossRef]

13. Orlowski, N.; Kraft, P.; Pferdmenges, J.; Breuer, L. Exploring water cycle dynamics by sampling multiple stable water isotope pools in a developed landscape in Germany. *Hydrol. Earth Syst. Sci.* **2016**, *20*, 3873–3894. [CrossRef]

14. Mook, W.; Gat, J.; Meijer, H.; Rozanski, K.; Froehlich, K. *Environmental Isotopes in the Hydrological Cycle: Principles and Applications—Vol. I, Introduction: Theory, Methods, Review*; IHP-V Technical Documents in Hydrology, N°39; UNESCO-IAEA: Vienna, Austria, 2001; pp. 61–70.

15. Jodar, J.; Custodio, E.; Liotta, M.; Lamban, L.J.; Herrera, C.; Martosrosillo, S.; Sapriza, G.; Rigo, T. Correlation of the seasonal isotopic amplitude of precipitation with annual evaporation and altitude in alpine regions. *Sci. Total Environ.* **2016**, *550*, 27–37. [CrossRef] [PubMed]

16. Cerar, S.; Mezga, K.; Žibret, G.; Urbanc, J.; Komac, M. Comparison of prediction methods for oxygen-18 isotope composition in shallow groundwater. *Sci. Total Environ.* **2018**, *631–632*, 358–368. [CrossRef] [PubMed]

17. Bortolami, G.C.; Ricci, B.; Susella, G.F.; Zuppi, G.M. *Isotope Hydrology of the Val Vorsaglia, Maritime Alps, Piedmont, Italy*; IAEA Proceedings Series v1; International Atomic Energy Agency: Vienna, Austria, 1979; pp. 327–378.

18. Darling, W.G.; Bath, A.H. A stable isotope study of recharge processes in the English Chalk. *J. Hydrol.* **1988**, *101*, 31–46. [CrossRef]

19. Yonge, C.J.; Goldenberg, L.; Krouse, H.R. An isotope study of water bodies along a traverse of southwestern Canada. *J. Hydrol.* **1989**, *106*, 245–255. [CrossRef]

20. Gieske, A.; Selaolo, E.; Mcmullan, S. Groundwater recharge through the unsaturated zone of southeastern Botswana: A study of chlorides and environmental isotopes. *IAHS-AISH Publ.* **1990**, *191*, 33–44.

21. Edirisinghe, E.A.N.V.; Karunarathne, G.R.R.; Samarakoon, A.S.M.N.B.; Pitawala, H.M.T.G.A.; Tilakarathna, I.A.N.D.P. Assessing causes of quality deterioration of groundwater in puttalam, Sri Lanka, using isotope and hydrochemical tools. *Isot. Environ. Health Stud.* **2016**, *52*, 1–16. [CrossRef]

22. Huang, T.; Pang, Z.; Li, J.; Xiang, Y.; Zhao, Z. Mapping groundwater renewability using age data in the Baiyang alluvial fan, NW China. *Hydrogeol. J.* **2017**, *25*, 743–755. [CrossRef]

23. Joshi, S.K.; Rai, S.P.; Sinha, R.; Gupta, S.; Shekhar, S. Tracing groundwater recharge sources in the northwestern Indian alluvial aquifer using water isotopes (δ^{18}O, δ^2H and ^{3}H). *J. Hydrol.* **2018**, *559*, 835–847. [CrossRef]

24. Scanlon, B.R.; Reedy, R.C.; Stonestrom, D.A.; Prudic, D.E.; Dennehy, K.F. Impact of land use and land cover change on groundwater recharge and quality in the southwestern US. *Glob. Chang. Biol.* **2005**, *11*, 1577–1593. [CrossRef]

25. Vrba, J.; Hirata, R.; Girman, J.; Haie, N.; Lipponen, A.; Neupane, B.; Shah, T.; Wallin, B. Groundwater resources sustainability indicators. In *The Global Importance of Groundwater in the 21st Century: Proceedings of the International Symposium on Groundwater Sustainability, Alicante, Spain, 24–27 January 2006*; Ragone, S., Ed.; National Groundwater Association: Westerville, OH, USA, 2007; pp. 129–138.

26. Huang, T.; Pang, Z. Estimating groundwater recharge following land-use change using chloride mass balance of soil profiles: A case study at Guyuan and Xifeng in the loess plateau of China. *Hydrogeol. J.* **2011**, *19*, 177–186. [CrossRef]

27. Vorosmarty, C.J.; Lettenmaier, D.P.; Leveque, C.; Meybeck, M.; Pahlwostl, C.; Alcamo, J.; Cosgrove, W.J.; Grassl, H.; Hoff, H.; Kabat, P.; et al. Humans transforming the global water system. *Eos Trans. Am. Geophys. Union* **2004**, *85*, 509–514. [CrossRef]

28. Mays, L.W. Groundwater resources sustainability: Past, present, and future. *Water Resour. Manag.* **2013**, *27*, 4409–4424. [CrossRef]

29. Allison, G.B.; Cook, P.G.; Barnett, S.; Walker, G.R.; Jolly, I.; Hughes, M.W. Land clearance and river salinisation in the western Murray basin, Australia. *J. Hydrol.* **1990**, *119*, 1–20. [CrossRef]

30. Liu, C.J.; Jiao, Y.M.; Zhang, G.L.; Liu, X.; Zhang, T.Y. Altitude effect of hydrogen and oxygen isotopes in spring water of the Malizhai river basin in Hani terrace. *J. Glaciol. Geocryol.* **2016**, *38*, 1404–1410.

31. Jiao, Y.M.; Liu, C.J.; Liu, X.; Liu, Z.L.; Ding, Y.P. Impacts of dominated landscape types on hydrogen and oxygen isotope effects of spring water in the Hani rice terraces. *Chin. J. Appl. Ecol.* **2017**, *28*, 2299–2306.

32. Yunnan Geological Survey Bureau. *1:200,000 Regional Geological Report of the Yuanyang Areas*; Yunnan Geological Survey Bureau: Kunming, China, 1971; pp. 20–31.

33. Jiao, Y.; Liu, C.; Gao, X.; Xu, Q.; Ding, Y.; Liu, Z. Impacts of moisture sources on the isotopic inverse altitude effect and amount of precipitation in the Hani rice terraces region of the Ailao mountains. *Sci. Total Environ.* **2019**, *687*, 470–478. [CrossRef]

34. Gourcy, L.L.; Groening, M.; Aggarwal, P.K. Stable oxygen and hydrogen isotopes in precipitation. In *Isotopes in the Water Cycle: Past, Present and Future of Developing Science*; Aggarwal, P.K., Gat, J.R., Froehlich, K.F.O., Eds.; Springer: Dordrecht, The Netherlands, 2005; pp. 39–51.

35. Jiao, Y.M. *Natural and Cultural Landscape Ecology of Hani Terraces Beijing*; China Environmental Science Press: Beijing, China, 2009; pp. 1–12.
36. Mezga, K.; Urbanc, J.; Cerar, S. The isotope altitude effect reflected in groundwater: A case study from Slovenia. *Isot. Environ. Health Stud.* **2014**, *50*, 33–51. [CrossRef]
37. Chamberlain, C.P.; Poage, M.A. Reconstructing the paleotopography of mountain belts from the isotopic composition of authigenic minerals. *Geology* **2000**, *28*, 115–118. [CrossRef]
38. Xing, X.H.; Liu, G.M.; Li, H.Q.; Hu, G.L. Sources of spring water and its characteristics of hydrogen and oxygen stable isotopes in Hasi Mountain. *J. China Hydrol.* **2016**, *36*, 46–50.
39. Dansgaared, W. Stable isotopes in precipitation. *Tellus* **1964**, *16*, 463–468.

MDPI

St. Alban-Anlage 66

4052 Basel

Switzerland

Tel. +41 61 683 77 34

Fax +41 61 302 89 18

www.mdpi.com

Water Editorial Office

E-mail: water@mdpi.com

www.mdpi.com/journal/water